Springer Theses

Recognizing Outstanding Ph.D. Research

For further volumes:
http://www.springer.com/series/8790

Aims and Scope

The series "Springer Theses" brings together a selection of the very best Ph.D. theses from around the world and across the physical sciences. Nominated and endorsed by two recognized specialists, each published volume has been selected for its scientific excellence and the high impact of its contents for the pertinent field of research. For greater accessibility to non-specialists, the published versions include an extended introduction, as well as a foreword by the student's supervisor explaining the special relevance of the work for the field. As a whole, the series will provide a valuable resource both for newcomers to the research fields described, and for other scientists seeking detailed background information on special questions. Finally, it provides an accredited documentation of the valuable contributions made by today's younger generation of scientists.

Theses are accepted into the series by invited nomination only and must fulfill all of the following criteria

- They must be written in good English.
- The topic should fall within the confines of Chemistry, Physics and related interdisciplinary fields such as Materials, Nanoscience, Chemical Engineering, Complex Systems and Biophysics.
- The work reported in the thesis must represent a significant scientific advance.
- If the thesis includes previously published material, permission to reproduce this must be gained from the respective copyright holder.
- They must have been examined and passed during the 12 months prior to nomination.
- Each thesis should include a foreword by the supervisor outlining the significance of its content.
- The theses should have a clearly defined structure including an introduction accessible to scientists not expert in that particular field.

David M. Kipping

The Transits of Extrasolar Planets with Moons

Doctoral Thesis accepted by
University College London, United Kingdom

Springer

Author
Dr. David M. Kipping
Rm. P-348, Center for Astrophysics
Harvard-Smithsonian Center
for Astrophysics
60 Garden Street
Cambridge, MA 02138
USA
e-mail: dkipping@cfa.harvard.edu

Supervisors
Prof. Dr. Alan Aylward
Department of Physics and Astronomy
University College London
London
UK
e-mail: alan@star.ucl.ac.uk

Prof. Dr. Giovanna Tinetti
Department of Physics and Astronomy
University College London
London
UK
e-mail: g.tinetti@ucl.ac.uk

ISSN 2190-5053
e-ISSN 2190-5061
ISBN 978-3-642-22268-9
e-ISBN 978-3-642-22269-6
DOI 10.1007/978-3-642-22269-6
Springer Heidelberg Dordrecht London New York

Cover design: eStudio Calamar, Berlin/Figueres

Printed on acid-free paper

Springer is part of Springer Science+Business Media (www.springer.com)

I dedicate this thesis to Emily, my morning light and brightest star

Supervisors' Foreword

Dave Kipping's thesis is remarkable in that it virtually single-handedly started a new field of study the search for exomoons, that is the moons of planets going around other stars. This started as an investigation of the statistical certainty one could place on transit observations. When Dave first looked at the mathematics of transit observations the field was using simple assumptions such as circular orbits: this thesis has illustrated that far more information is available about exoplanet orbits, and in particular potentially about exomoons, than was conceived of three or four years ago.

This work is significant because it has been at the vanguard of the development of the statistical theory of transit observations. Dave has himself been involved in observations, initially at UCL but later using more impressive facilities, and has become through his work an accepted member—often a leader—in observational campaigns using both ground- and space-based facilities.

His promotion of the idea that exomoons may be one of the most important areas to look for habitable worlds has brought him to the attention to a wider community. His work now forms a significant part of the search for exo-objects with facilities such as the Kepler probe, and has had a significant impact in outreach to the wider public, for whom the idea of habitable moons has an exciting edge.

From his thesis work Dave has gone on to an important position at CfA in Harvard which gives him access to Kepler data and the chance to put his ideas into action.

London, April 2011

Dr. Alan Aylward
Dr. Giovanna Tinetti

Preface

The search for extrasolar planets is strongly motivated by the goal of determining how frequent habitable worlds and life may be within the Galaxy. Whilst much effort has been spent on searching for Earth-like planets, large moons may also be common, temperate abodes for life. The methods to detect extrasolar moons, or "exomoons" are more subtle than their planetary counterparts and in this thesis I aim to provide a method to find such bodies in transiting systems, which offer the greatest potential for detection.

Before one can search for the tiny perturbations to the planetary signal, an understanding of the planetary transit must be established. Therefore, in Chaps. 3 to 5 I discuss the transit model and provide several new insights. Chapter 4 presents new analytic expressions for the times of transit minima and the transit duration, which will be critical in the later search for exomoons. Chapter 5 discusses two sources of distortion to the transit signal, namely blending (with a focus on the previously unconsidered self-blending scenario) and light curve smearing due to long integration times. I provide methods to compensate for both of these effects, thus permitting for the accurate modelling of the planetary transit light curve.

In Chap. 6, I discuss methods to detect exomoons through their gravitational influence on the host planet, giving rise to transit timing and duration variations (TTV and TDV). The previously known TTV effect is updated with a new model and the associated critical problems are outlined. I then predict a new effect, TDV, which solves these problems, making exomoon detection viable. Chapter 7 presents a feasibility study for detecting habitable-zone exomoons with Kepler, where it is found that moons down to $0.2M_{\oplus}$ are detectable. Finally, conclusions and future work are discussed in Chap. 8.

Acknowledgments

I am immensely grateful to those who have supported me during both my Ph.D. studies and those who guided and nurtured my insatiable curiosity about the natural world since my earliest memories.

Firstly, I thank my official supervisors, Giovanna Tinetti and Alan Aylward, for their unerring support and guidance throughout the last three and a half years. I feel fortunate to have had supervisors who afforded me the freedom to pursue my own ideas and encouraged such thinking throughout. Special thanks to Jonathan Tennyson and Hugh Jones, my thesis examiners, for taking the time to so carefully read this work.

I owe an enormous debt of gratitude to Giovanna Tinetti who always found time for me and provided a level of support which has been the envy of other graduate students. I feel privileged to have been the student of such a talented scientist and I extend my warmest thanks to her for everything she has done for me.

I offer a special thanks to Gáspár Bakos, who invited me to the CfA as a predoc. I never met anyone who has such a formidable knowledge of so many different aspects of the art of detecting exoplanets and it was a privilege to have worked with Gáspár. I would also like to acknowledge the continual support and friendship of Jean-Philippe Beaulieu who has made me smile many times during those darker days of a graduate student's life.

I am grateful to the teachers of Twycross House School who taught me how to apply and focus my mind in a way which changed the path of my life. In particular, I thank Michael Fox who nurtured my passion for physics and the Kirkpatricks for their patience and kindness. During my time at Cambridge, I would like to thank Bill Nolan for his guidance and support. Additionally, I am grateful to Roland Gähler for inviting me to work at L'Institut Laue Langevin and his immense hospitality during that time.

I would like to thank my mother and father for their love and immeasurable support throughout my life. Without them, none of my achievements would have been possible. I also thank my sister, Louisa, and grandparents, Reneé and Alec, for all they have done for me. Finally, I thank Emily, the sunshine of my life, who makes me feel like I can achieve anything. I cannot find words to express my love for you.

Contents

Chapter 1
Introduction

To consider the Earth as the only populated world in infinite space is as absurd as to assert that in an entire field sown with millet, only one grain will grow

Metrodorus of Chios, fourth century BC

1.1 Are We Alone?

It is no exaggeration to say that a great deal of the motivation behind searching for extrasolar planets is embodied by the above question. Since time immemorial, humans have turned their gaze towards the vast, illimitable, velvet darkness peppered with oases of light and wondered whether other beings, such as ourselves, also populate the ocean of space and time.

With the Earth representing the only environment where life is known to reside, a natural place to begin a search for extraterrestrial life are environments similar to this sole example. Consequently, planets, and in particular "Earth-like" planets, are believed to offer the best locations for life to begin. Just like the ancient fascination with extraterrestrial life, planets have long captured the imagination of human minds.

Since antiquity, the seven *asteres planetai*, or the "wandering stars", were known to many cultures across the world. These bodies, all visible to the naked eye, were noted as abnormal heavenly objects in that they did not appear in the same part of the sky each night. The seven are the Sun, the Moon, Mercury, Venus, Mars, Jupiter

D. M. Kipping, *The Transits of Extrasolar Planets with Moons*, Springer Theses,
DOI: 10.1007/978-3-642-22269-6_1, © Springer-Verlag Berlin Heidelberg 2011

and Saturn, which give us the seven days of the week. It was the irregular *motions* of these objects which distinguished them as planets and this provides a natural place for us to start our exploration too.

1.2 Motions of the Planets

1.2.1 Celestial Mechanics

Whilst the positions of the planets were rigorously recorded and tracked by the Babylonians as early as the second millennium BC, it was the Greeks who introduced a geometric model of their motions. A geocentric view of the Universe was employed, presumably since it was natural to assume the Earth was at rest, leading to elaborate models to describe the planetary motions. The pinnacle of these theories was presented in *Almagest* by Ptolemy [1] in the second century AD utilizing a complex arrangement of planetary spheres and epicycles. Ptolemy's theory was so successful at predicting these motions it remained the definitive text for thirteen centuries.

It was not until the European renaissance that the heliocentric model of Copernicus, Galileo and Kepler gained favour in the sixteenth century and the definition of a planet changed from something which orbited the Earth to something which orbited the Sun, demoting the number of planets from seven to six.[1] Galileo Galilei's telescope design in 1609 paved the way for the future discovery of Uranus [2–4] and Pluto in 1930 Tombaugh discovered (as well as the moons of these planets), ushering in the modern view of the Solar System.

The other revolution came with Sir Isaac Newton's *Philosophiae Naturalis Principia Mathematica*, first published on the 5th July 1687 [5]. Whilst Johannes Kepler had earlier provided his three laws of planetary motion which so accurately predicted the motions of the planets, the laws were simply empirical observations without a fundamental understanding as to why the three laws worked. Newton's theory showed that planets move in ellipses due to a single force, the force of gravity, and that a simple inverse-square force potential could completely explain all of the observed motions of the planetary bodies.

1.2.2 Einstein's General Theory of Relativity

Despite the many great successes of the Newtonian theory, Le Verrier et al. [3] reported that the perihelion precession of Mercury's orbit around the Sun could

[1] Although the Moon and the Sun were no longer classed as planets, the Earth was now seen as a planet, to give six

not be completely explained with Newton's theory. Mercury's precession is measured to be 5600 as/century whereas Newtonian mechanics predicts a precession of 5557 as/century [6]. At the time, Le Verrier proposed an additional planet named "Vulcan", which he suggested resided between Mercury and the Sun.

However, the discrepancy was later resolved without the need to invoke another planet when Albert Einstein published his General Theory of Relativity in 1915 [7]. The observed motions of the planetary bodies was finally completely understood by the early twentieth century.

1.2.3 The Discovery of Neptune

The discovery of Neptune in many ways serves as a transitional stage between the methods used to detect Solar System planets and the methods later used to detect extrasolar planets. Bouvard [8] first struck upon the hypothesis of Neptune's existence when he tabulated the observations of the orbit of Uranus and noticed substantial deviations from current predictions. Bouvard hypothesized that an unknown outer planet was perturbing the orbit of Uranus. Le Verrier et al. [3] and Adams [9] went further by calculating what the orbit of this body must be. These predictions were sufficiently accurate that the detection of Neptune was made soon after by Galle [4].

Neptune therefore represented a transitional stage between the classic technique of directly observing the reflected starlight from planets and moons, to detecting bodies through indirect techniques based upon the predictions of celestial mechanics. Celestial mechanics had matured to a sufficient point that it could be a tool for planetary detection, which would later play a vital role in exoplanetary searches.

1.3 Historical Overview of Efforts to Find Extrasolar Planets

1.3.1 From Planets to Exoplanets

After the detection of Pluto in 1930 and the picture of the Solar System apparently complete, many astronomers set their sights on much more distant planets—those orbiting other stars. This ambitious goal would not be achievable through the conventional techniques used to find Solar System planets.

These planets, including the seven *asteres planetai*, were found by detecting the sunlight reflected off their surface as they moved against the backdrop of a black canvas. Therefore, the oldest technique used to find planets is what we would now call "direct imaging".

In contrast, the light reflected from an extrasolar planet, or "exoplanet", is very challenging to spatially resolve from the host star and is typically around one million to one billion times fainter (see Sect. 1.3.8). In recognition of this, astronomers had to devise new indirect methods in their bid to detect an extrasolar planet.

1.3.2 Orbital Perturbations

As discussed earlier in the case of Neptune, the second technique to yield success in detecting Solar System planets was by searching for deviations in the motions of a known orbiting body, betraying the presence of a unseen perturber. The use of this technique reflects the extreme confidence astronomers now had in their understanding of the motions of heavenly bodies, in large part due to the pioneering work of Kepler, Newton, Lagrange, Hill, Einstein and many others.

These orbital deviations may be classed into two categories depending upon the timescale over which they are observable. The first category is orbit-to-orbit variations (e.g. Le Verrier's prediction of Neptune) and the second is long-term secular (or apparently so) changes in the orbital elements of the system (e.g. the prediction of Vulcan, also by Le Verrier).

In both cases, one requires at least two bodies in the system to detect the third. Naturally, this limits the use of the technique to either a multiple-star system or a star for which exoplanets are already known. In the nineteenth century, the latter of these was of course not available. Therefore, following the success of Le Verrier's work with Neptune and the known existence of numerous bright multiple star systems, it should perhaps come as no surprise that one of the first scientific claims of an exoplanet detection was for a binary star system.

In 1855, Capt. W. S. Jacob at the East India Company's Madras Observatory reported that orbital anomalies in the binary star 70 Ophiuchi made it "highly probable" that there was a "planetary body" in this system. Jacob [10] compared observations of the positions of the binary components to a calculated model of a two-body and three-body system and noted that the average residuals decreased from 49 to 35 amin over the entire data set (see Fig. 1.1). The statement "highly probable" was not quantified in the way which would be mandatory by modern standards, but Jacob can be forgiven, given that much of seminal work in observational statistics by Karl Pearson (Pearson's chi squared test in 1900 [11]), Ronald Fischer (the F-test in 1920s) and others had not been devised yet. Using Jacob's original data, a modern F-test would find the three-body model to be accepted over the two-body with a confidence of 1.9-σ, which is below the oft-cited detection threshold of 3-σ.

Many interesting aspects of nineteenth century astronomy are visible when one reads through Jacob's paper with a modern perspective. For example, Jacob remarks that the deviations "might arise from an erroneous assumption of the universal application of the law of gravitation". Jacob essentially concludes that only two hypotheses can explain the observations; either Newton's laws of gravity requires modification or there is a planet in orbit. Between these two hypotheses, Jacob quite reasonably concludes the latter is much more likely. There are no other hypotheses put forward; the idea that instrumental or human errors could be responsible for

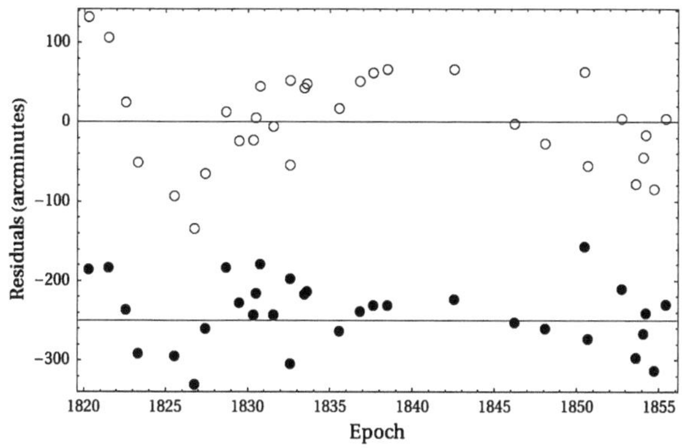

Fig. 1.1 Original 1855 data of Capt. W. S. Jacob, comparing the residuals from a two-body model (*open circles*) for 70 Ophiuchi and a three-body model (*filled circles*, offset by −250 arcminutes for clarity). The decreased residuals of the three-body model was used to make one of the first claims for the existence of a exoplanet. A modern statistical analysis using an F-test finds the three-body model to be accepted with 1.9-σ confidence, below the standard detection threshold

the observations is notably absent. By virtue of the fact this third hypothesis is not even mentioned, it therefore follows that Capt. Jacob considered a systematic error much less likely than his second hypothesis—that the laws of gravitation are not universal—a concept which would be viewed as grossly aberrant to the modern exoplanetary scientist.

A few decades later, See [12] made a stronger claim for the existence of a dark companion in this system. However, Forest Ray Moulton, who had been a graduate student of See at the University of Chicago, soon published a paper proving that a three-body system with the specified orbital parameters would be highly unstable [13]. The claims by Jacob and See have both been shown to be erroneous [14].

1.3.3 Astrometry

One of the oldest and most famous scientific techniques used to look for an exoplanet is that of astrometry. Unlike the orbital perturbations method, there is no example of a Solar System planet ever being discovered through astrometry and so in many ways astrometry was quite a radical technique. What is astrometry and why was it proposed so early on without any successes in the Solar System?

From Newton's theory, two masses in an inertial frame orbit a common centre of mass. This means that a planet doesn't just have an orbit, the star also has an orbit around the centre-of-mass, albeit of much smaller amplitude. Figure 1.2 provides an illustration of this and [15] give a detailed discussion of the two-body problem. The star's "reflex motion" to the orbiting mass could therefore potentially be used to detect an exoplanet, if stellar positions could be measured to sufficient precision.

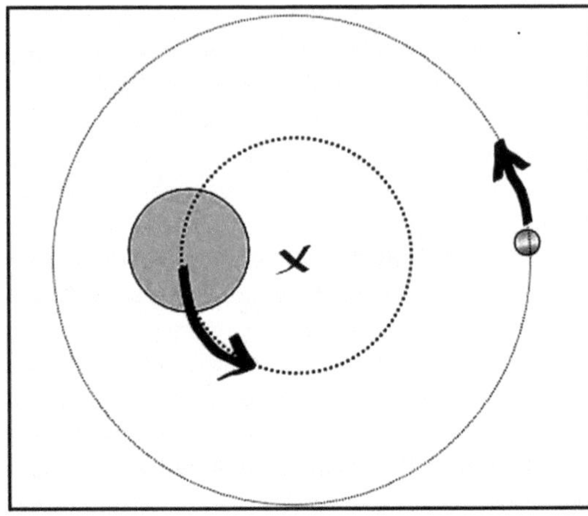

Fig. 1.2 Cartoon illustrating the reflex motion of a star in the inertial frame of the barycentre. The planet's presence gives rise to variations in the position (used for astrometry and pulsar timing) and the velocity (used for radial velocity) of the host star, which may used to detect an exoplanet

The variation in the sky position, α, of a star of mass M_* at a distance d from the Earth, hosting a planet[2] of mass M_P and semi-major axis a_P, can be estimated by Eq. (1.1). The equation reveals that $\sim 10^{-7}$ as, or ~ 0.1 μas is a reasonable estimate for the required precision in astrometry, which outlines just how challenging detecting an exoplanet truly is with this method.

$$\begin{aligned} M_P a_P &= M_* a_* \\ \alpha &\simeq \frac{a_*}{d} \\ &= (94.''5 \times 10^{-9}) \left(\frac{a_P}{\text{AU}}\right) \left(\frac{10\ \text{pc}}{d}\right) \left(\frac{M_P}{M_J}\right) \left(\frac{M_\odot}{M_*}\right) \end{aligned} \tag{1.1}$$

Where "AU" denotes astronomical units, "pc" denotes parsec, M_J is the mass of Jupiter, $M_\oplus$ the Solar mass unit and a_* represents the semi-major axis of the star's reflex motion.

At the time when astrometry was first implemented, it was of course not the only considered method. For example, transits of Venus and Mercury had long been known. The transit of Venus was first observed in 1032 by the Persian astronomer Avicenna, who used the observation to conclude that Venus is closer to the Earth than the Sun. In fact, one can go back to even the work of Ptolemy, who wrote about the possibility of planetary transits across the face of the Sun in his work "Planetary Hypotheses". He suggested that no transits had been observed either because planets such as Mercury were too small to see, or because the transits were too infrequent.

Despite the known existence of planetary eclipses, transits would not have seemed an appealing method of detecting exoplanets in the nineteenth and early

[2] I here introduce the notation for several parameters related to the planet and star. In many other texts, a subscript "P" is not placed on the planetary terms. However, in this thesis I will be dealing with >1 planet and satellites and the advantage of the "P" subscript will become apparent.

twentieth century. For a Jupiter-sized planet transiting a Sun-like star, the flux decrement would be $\sim$1% (see Sect. 1.3.6). Using just the human eye or even later with photographic plates, this level of precision was simply not feasible. However, astronomers were very skilled in measuring positions of objects in the sky since the late eighteenth century and early nineteenth century, able to reach better than an arcsecond precision, through improved telescope designs (for example [16]) measured the parallax of the binary star 61 Cygni to be 300 mas[3]).

Other reasons why transits were likely seen as unfeasible, even in the age of photomultipliers after 1934, was that transits are low probability events, both geometrically and temporally, meaning that a very large dedicated survey would have been required at great expense by nineteenth and early twentieth century standards. Finally, the expectation was to find planets on large distant orbits, based upon the configuration of the Solar System, which maximizes the astrometric amplitude[4] and minimizes the geometric probability of a transit.

Therefore, astrometry can be seen to have been destined to emerge as one of the first methods to look for exoplanets. Observers were no doubt spurred on by the early successes of Friedrich Bessel. Bessel [17] used astrometry to detect the presence of unseen dark companions to Sirius and Procyon (astrometric amplitudes of 3.7 mas and 1.8 mas respectively), although these were unseen stars rather than exoplanets. It is worth pointing out that these detections did withstand the scrutiny of later observations.

Over a hundred years after Bessel's groundbreaking work in astrometry, the stage was set for one of the most famous stories in the tale of exoplanets. In 1963, Peter van de Kamp claimed to have detected an exoplanet orbiting a nearby M-dwarf known as "Barnard's star" (named after the discoverer of its very high proper motion). Van de Kamp [18] argued for a 1.6 Jupiter mass (M_J) planet on a 4.4 AU orbit electrifying the astronomical community (see Fig. 1.3 for the original data).

In a familiar pattern though, the signal was soon identified as spurious. In subsequent observations, Gatewood and Eichhorn [19] failed to verify the planetary signal and around the same time Herschey [20] showed that that changes in the astrometric field of various stars correlated to the timing of adjustments and modifications that had been carried out on the refractor telescope's objective lens. As a result, the planetary "discovery" was an artefact of maintenance and upgrade work on the telescope and is an excellent example of an uncorrected systematic effect masquerading as a planet.

Barnard's star may have seemed an attractive target as it has a low mass, being an M-dwarf, and is one of the nearest stars to the Solar System meaning the claimed signal of Van de Kamp was (24.5 ± 2.0) mas. However, Barnard's star also has the highest proper motion out of any star in the sky at a formidable 10.3 as/year. This means over the putative 24 year period of Van de Kamp's planet, the star exhibits a systematic motion over 10,000 times larger than the signal due to the proposed planet.

[3] Note: "mas" denotes milli-arcseconds.

[4] And also minimizes the radial velocity amplitude.

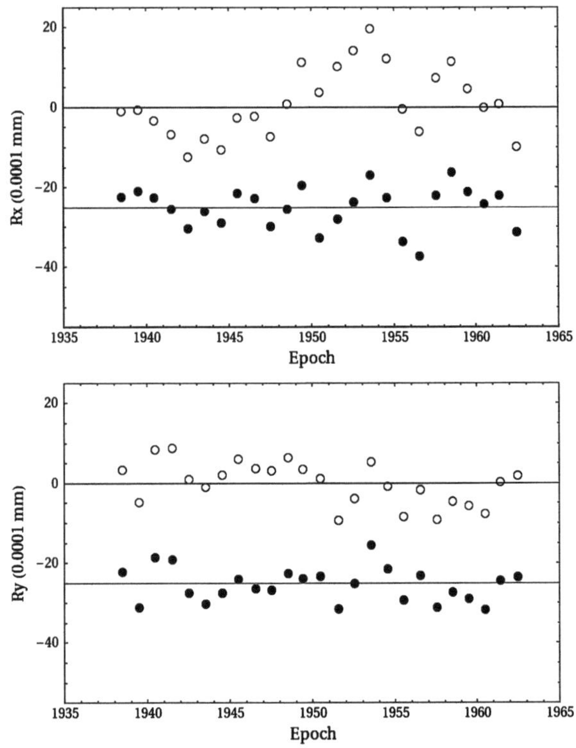

Fig. 1.3 Original 1963 data of Peter van de Kamp, comparing the residuals from a two-body model (*open circles*) for Barnard's star and a three-body model (*filled circles*, offset by $-25\,\mu$m for clarity). The decreased residuals of the three-body model was used to claim the presence of a 1.6 M_J exoplanet in a 4.4 AU orbit. A modern statistical analysis using an F-test finds the three-body model to be accepted with 1.7-σ confidence, below the standard detection threshold

It is worth noting that despite the early problems with astrometry, it has since been exploited on several targets known to already host planets for the purpose of characterizing the orbit and demonstrating the feasibility of astrometry in general. An excellent example comes from Benedict et al. [21], who use HST astrometry to determine the mass of GJ 876b.

1.3.4 Pulsar Timing

After the debacle over Barnard's star, it is easy to see how confidence may been lost in the hunt for exoplanets. By the beginning of 1990, over 130 years had passed of claimed detections of exoplanets which were consistently debunked by later studies.

In the same decade that Van de Kamp wrote his infamous paper on Barnard's star, Hewish et al. [50] detected the first pulsar (a highly magnetized rotating neutron star) which they designated "Little Green Men 1" or LGM01 (now less poetically known as PSR B1919+21). The original designation came from the half tongue-in-cheek proposition that the measured regular radio beats were artificial in nature, as there was previously no concept of a rotating star beaming electromagnetic radiation in this manner. The few dozen pulsars subsequently discovered were subject to frequent timing measurements, in order to pin down their positions and spin-down parameters.

The 6.2 ms pulsar PSR B1257+12 became an object of interest in the early 1990s due to irregularities in the timing of the pulses, as measured by the team of Aleksander Wolszczan. The object was studied further by the giant 305 m Arecibo telescope by the same group and led to the inescapable conclusion that deviations in the times-of-arrival (TOA) of the pulsar must be due to the presence of two perturbing planets. Wolszczan and Frail [22] published their results soon after in what would become historically acknowledged as the first unambiguous detection of an extrasolar planet.

On a brief aside, Wolszczan and Frail [22] were under the impression that their detection was not the first at the time of writing. A year earlier, Bailes et al. [23] published a letter to Nature claiming a 10 $M_{\oplus}$ planet orbiting the pulsar PSR B1829−10 with an orbital period of nearly exactly 6 months. The authors later retracted the planet upon realising they failed to properly account for the Earth's elliptical orbit.

The method of looking for TOA deviations is very similar to astrometry. Due to the changing position of the star, as a result of its reflex motion, the distance between the observer and the pulsar changes periodically. These changes in distance result in a shorter and longer light travel times.

$$
\begin{aligned}
\Delta(\mathrm{TOA}) &\simeq \frac{a_*}{c} \sin i_P \\
&= \frac{a_P M_P}{M_* c} \sin i_P \\
&= (4.74 \times 10^{-4}\ \mathrm{s}) \sin i_P \left(\frac{a_P}{\mathrm{AU}}\right) \left(\frac{M_P}{M_J}\right) \left(\frac{M_\odot}{M_*}\right)
\end{aligned}
\tag{1.2}
$$

Where i_P is the orbital inclination of the planet. Equation (1.2) shows a typical signal is around half a millisecond which was much larger than the typical uncertainties of a few hundred nanoseconds being achieved for millisecond pulsars by the time of 1990. As a result, the signal seen by Wolszczan and Frail [22], which I show in Fig. 1.4, was extremely secure from a statistical perspective.

Although the significance was not quoted in the original paper, I have estimated this by digitizing the figure of the orbital fit and comparing the root mean square (RMS) between the two-planet model and the null hypothesis of a static system. The result is a false-alarm-probability (FAP) of 4.3×10^{-40} or 13.3-σ, which gives a sense of just how secure the detection was relative to that of Jacob [10] and Van de Kamp [18]. Note that Wolszczan and Frail [22] also considered whether their signal could be a systematic error by verifying that the observed periodicities were independent of radio frequency and that other millisecond pulsars routinely observed with the same equipment did not show such periodicities either.

The Wolszczan and Frail [22] paper is recognized as the first unambiguous detection of a planet orbiting another star. What made the paper even more remarkable was that the planets detected were in the terrestrial mass regime at 2.8 $M_{\oplus}$ and 3.4 $M_{\oplus}$ and what would be considered "long-period" from a modern perspective, with respective orbital periods of 98.2 and 66.6 days.

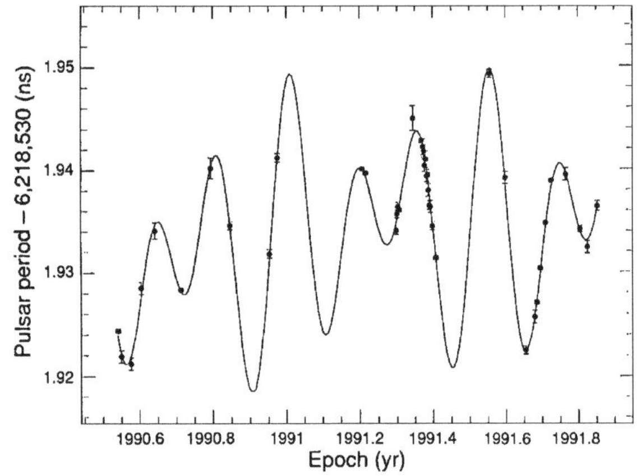

Fig. 1.4 Original data of Wolszczan and Frail [22], showing a two-planet fit to the measured pulsar times of PSR B1257+12. In contrast the earlier claims of Jacob [10] (see Fig. 1.1) and Van de Kamp [18] (see Fig. 1.3), the signals of each planet are overwhelming obvious and have a significance of roughly 13-σ. Figure taken from Wolszczan and Frail [22]

However, despite the low masses and long periods, pulsars would be a hellish environment for life and there was still a strong motivation to find a planet around a "normal" star, i.e. main-sequence. In many ways, the pulsar detection spurred on the exoplanet hunt tremendously since now astronomers knew stars hosted planets and yet the real prize of a planet around a Sun-like star remained undiscovered.

1.3.5 Radial Velocity

1.3.5.1 Principles of RV

In many ways, the age of exoplanet discovery did not truly begin until the detection of 51-Pegasi b, discovered using the radial velocity technique. Radial velocity, or RV, is very similar to the astrometric method except that instead of looking for the reflex motion of the star in terms of changes in position, one looks for changes in velocity (along the line of sight). Since a planet, and thus the star's corresponding reflex motion, orbit with a higher speed when in tighter orbits, RV favours planets on very short orbital periods. This is in contrast to astrometry which favours long-period planets yielding a large planet-star separation and thus large amplitude spatial motion of the host star. I summarise these very important points below.

- Astrometry: Measures the sky-projected **position** of a star: Prefers long-period companions
- Radial Velocity: Measures the line-of-sight **velocity** of a star: Prefers short-period companions

Therefore, RV and astrometry are highly complementary to one another in parameter space. Additionally, the combination of the two allows for a unique solution for the planetary mass [24]. This complementarity between a position-based phenomenon and a velocity-based phenomenon will have important ramifications later when I discuss detecting exomoons and so I stress the point early on. The reflex velocity semi-amplitude can be easily written down for a circular orbit to give a feeling as to the feasibility of such an enterprise. One may again exploit the fact that the planetary motion and the stellar motion must balance out in the inertial frame:

$$\begin{aligned} M_P v_P &= M_* v_* \\ K_* \simeq v_* \sin i_P &= \frac{M_P \sin i_P}{M_*} \frac{2\pi a_P}{P_P} \\ &= M_P (2\pi G)^{1/3} M_*^{-2/3} P_P^{-1/3} \\ &= (28.4 \text{ m/s}) \sin i_P \left(\frac{M_P}{M_J}\right) \left(\frac{M_*}{M_\odot}\right)^{-2/3} \left(\frac{P_P}{\text{years}}\right)^{-1/3} \end{aligned} \tag{1.3}$$

On the second line, I have used K_* to denote the RV semi-amplitude, as is standard in the exoplanet literature. On the third line, I have used Newton's version of Kepler's Third Law to remove the dependency on a_P. Given that by 1995 typical RV errors could reach 20 m/s, Eq. (1.3) reveals that for giant planets on an orbit of a year or less, RV would be a viable detection technique.

The RV method requires measurements of a star's velocity along the line-of-sight, which can be achieved by measuring the Doppler shifts of said star's spectral lines. This naturally requires a very stable spectrograph, with highly sensitive calibration and a rich forest of lines to measure. To make the method feasible for an exoplanet search, the process has to become somehow automated to expedite what would otherwise be a very tedious task by hand (as well as prone to inaccuracies).

1.3.5.2 A Shot in the Dark

For RV to succeed, the key hinge was whether gas giants would exist in orbits of a year or less and thus be detectable. Planetary formation models [25] of the Solar System perpetrated that gas giants formed beyond the snow-line (where the equilibrium temperature is less than 150K which means 2–3 AU typically). Whilst inward migration was conceived, inspection of the positions of the Solar System gas giants alluded that this mechanism was not very effective. Thus, even after the Wolszczan and Frail [22] discovery, little resources were devoted to looking for exoplanets using the RV method. Nevertheless, some groups did take the gamble, which ultimately paid off for Mayor and Queloz [26].

Before discussing the famous paper of Mayor and Queloz [26], I point out that some tentative successes had already transpired for the RV method before 1995. Notably, seven years prior, Campbell [27] cautiously claimed to have detected a companion to Gamma Cephei which was challenged a few years later by Walker [28]. It was therefore not generally considered a confirmed planet until validated in Hatzes et al. [29]. Another example comes from Latham et al. [30] who detected a companion of minimum mass $11M_J$ around HD 114762, but concluded the object was more likely to be a brown dwarf than a planet.

The story of Mayor and Queloz [26] centres around "51-Peg"; a bright ($V = 5.5$) G2- to G4-type star relatively close at 15.6 pc in the Pegasus constellation. The star was part of a exploratory survey being conducted by a Swiss group based in Geneva, led by Michel Mayor looking at 142 bright K and G dwarfs for radial velocity variations with a sensitivity of $\sim$13 m/s.

Although not quoted in the original paper, it is simple to download the original data, refit a circular orbit model (no linear drift) and compute the significance of the fit. The result is a false alarm probability (FAP) of 7.1×10^{-19} or 8.9-σ, easily above the standard detection threshold of 3-σ. It is also visible that Mayor and Queloz [26] spend considerable effort excluding alternate hypotheses for the signal, for example from a blended eclipsing binary by virtue of a bisector analysis.

1.3.5.3 An Explosion of Discoveries

Just six days after the Mayor and Queloz [26] discovery, Geoff Marcy and Paul Butler of San Francisco State University independently confirmed the signal and then charged on with two new discoveries for 70 Virginis [31] and 47 Ursae Majoris [32]. And so began the age of exoplanetary science with new discoveries coming in on a regular basis from here on. Today, over 500 exoplanets are known to exist with over 80% coming from RV surveys.

Due to the bias of RV surveys to high mass, short-period planets, a large fraction of these bodies are previously unanticipated "hot-Jupiters". The fact that hot-Jupiters exist was a major boon for RV surveys in the early years, but modern high resolution spectroscopy can now measure stellar velocities to better than 1 m/s meaning planets down to a few Earth-masses (Super-Earths) at long periods can be found [33].

One interesting consequence of the RV discoveries was that many of the exoplanets being found had unanticipated short-periods (a few days) and thus had a much higher chance ($\sim$10%) of transiting the face of the star than previously expected. After the paper of Mayor and Queloz [26], astronomers started to think about the technique which had long been considered implausible (Fig. 1.5).

1.3.6 Transiting Planets

Despite RV surveys churning out plenty of exoplanet discoveries, RV provides frustratingly little information about these bodies. For example, the true mass of the planet is not actually known, just $M_P \sin i_P$. The only other information one can glean is the orbital period, P_P, the orbital eccentricity, e_P and the argument of periastron, ω_P. For the ultimate goal of detecting signs of life on another planet, this was simply not enough. If one of the RV detected planets happened to transit across the star though, vast amounts of detail could be inferred about the planet's

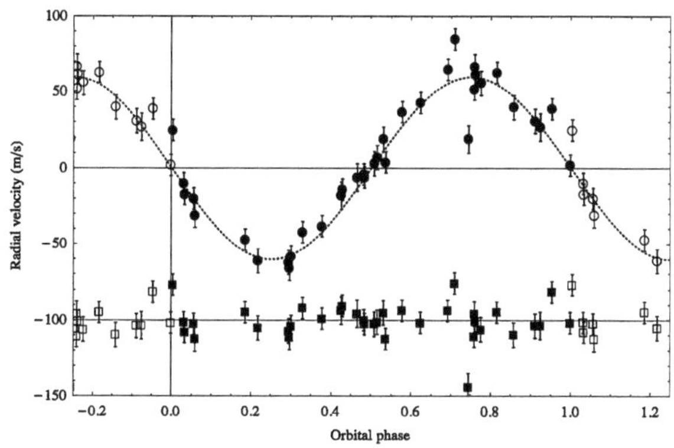

Fig. 1.5 Original data of Mayor and Queloz [26] (*circles*), showing a one-planet fit (*continuous dotted line*) to the measured radial velocities of 51-Pegasi. In contrast to the earlier claims of Jacob [10] (see Fig. 1.1) and Van de Kamp [18] (see Fig. 1.3), the signal of the planet is very clear and has a significance of roughly 9-σ. Squares show the offset ($-100\,\mathrm{m/s}$) residuals of the one planet circular-orbit fit. Empty points represent a repetition of the data to illustrate to signal more clearly

size, composition, oblateness, atmosphere, temperature, albedo, surface coverage and ultimately habitability.[5] Transits are the window into the soul of an exoplanet.

The principles of the transit method are very simple at a first-order level. The star has a nominal flux level which temporarily decreases due the planet blocking out a fraction of the projected stellar surface as it passes in front. The depth of the transit[6] is given by the ratio of the sky-projected area of the planet and the sky-projected area of the star:

$$\begin{aligned} \delta &= \frac{\pi R_P^2}{\pi R_*^2} \\ &= 1.03 \left(\frac{R_P}{R_J}\right) \left(\frac{R_\odot}{R_*}\right) \% \end{aligned} \tag{1.4}$$

By the end of the twentieth century, even small telescopes (a few inches aperture) had photometric precisions better than 1%, in large part due to the development of CCDs a few decades earlier. The key question was how likely was it that a transit would occur, given the plainly fortuitous geometric alignment required?

Assuming a planet is on a nearly circular orbit and the inclination angle is uniformly distributed in space, the probability is simply given by $\sim R_*/a$. For the hot-Jupiters being found from RVs, a good rough estimate was 10% probability. In reality, it is larger than this since RV surveys have a bias to aligned systems, but this conservative estimate is already encouraging. Consequently, success was soon coming.

[5] Note that not all of these possibilities were known early on.

[6] I here describe the geometric transit depth. The observed transit depth is, in general, deeper than this due to limb darkening effects.

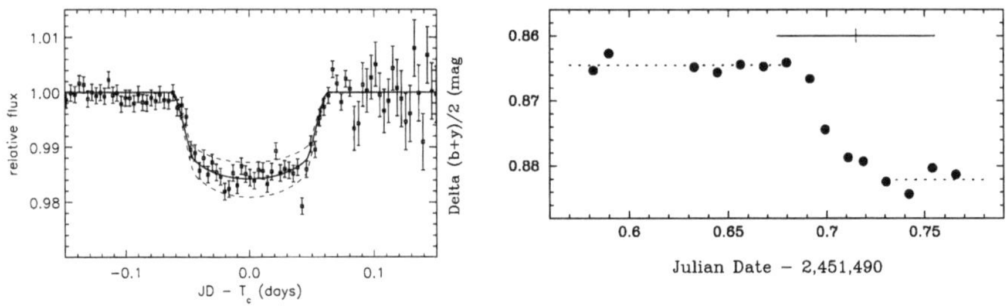

Fig. 1.6 First detected transit light curve of an exoplanet, made by Charbonneau et al. [34] (*left*) and Bakos et al. [36] (*right*) for HD 209458b. On the *left*, the *solid line* shows the best-fit model and the *dotted line* shows the extrema of a planet which is 10% smaller or larger in radius. *Left* figure taken from Charbonneau et al. [34]. *Right* figure taken from Bakos et al. [36]

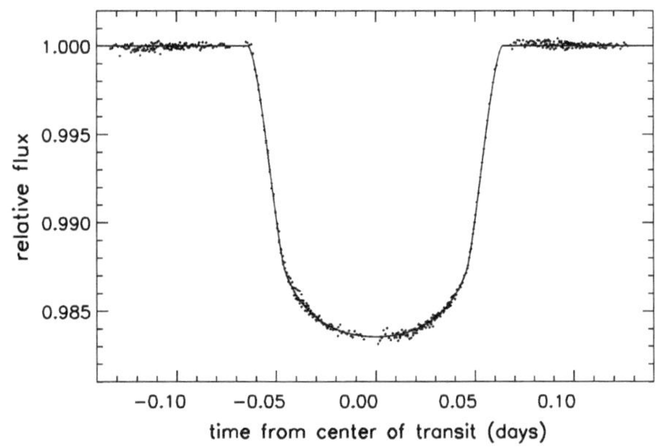

Fig. 1.7 First space-based transit light curve of an exoplanet, made by Brown [38] for HD 209458b using the Hubble Space Telescope (HST). Data has been phase-folded from five partial transits. It was not possible to observe a complete and uninterrupted transit, due to the 90 min orbit of HST. Figure taken from Brown [38]

Charbonneau et al. [34] and Henry et al. [35] detected the first transit of an extrasolar planet for HD 209458b, a fairly typical hot-Jupiter around a bright star (see Fig. 1.6). Charbonneau et al. [34] used the 4-inch STARE telescope, demonstrating just how inexpensive transit observations could potentially be (which opened the door for future ground-based transit surveys such as HATNet [36] and WASP [37]). The resulting light curve showed a depth of around 1.5% lasting for around 3 h. The light curve allowed for a determination of the orbital inclination angle and thus the true planetary mass from the RV determined $M_P \sin i_P$. Also, since the transit depth is essentially $(R_P/R_*)^2$, the planetary radius could be calculated and thus the average density of HD 209458b was known. The power of transits had been demonstrated.

A spectacular follow-up paper by Brown [38] used the STIS instrument on HST to provide the first space-based transit light curve. The stunning photometry, shown in Fig. 1.7, provided an exquisite determination of the system parameters and toyed with the idea of constraining the presence of rings and moons.

Having finally reached the discovery of transiting exoplanets, which constitutes the major topic of interest of this thesis, the relevant history of exoplanets has been

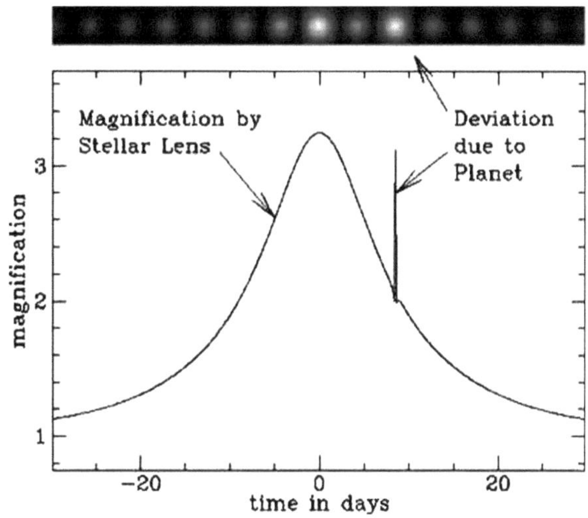

Fig. 1.8 Example microlensing light curve of a star with a planet. One can see the brightening due to the lensing of the star and later the effect of the planet is visible. Image taken from the Microlensing Planet Search Project homepage (http://bustard.phys.nd.edu/MPS/)

covered, for the purposes of this thesis. However, for completion, I will briefly overview two other known successful methods to detect exoplanets. Methods based upon transit timing will be discussed later in the thesis, in Chap. 6.

1.3.7 Gravitational Microlensing

Gravitational microlensing is based upon the fact that massive bodies bend the apparent path of light, in essence acting as a lens. If a star passes in front of another more distant luminous body, it causes the luminous body to dramatically increase in brightness for a few days or weeks, depending on the configuration. The same is true for a star with a planet except that one sees two (or more) increases in brightness; one large increase due to the more massive star and one smaller increase due to the planet (see Fig. 1.8 for an illustration).

Microlensing favours massive planets with a wide separation leading to the detection of cold or frozen planets. However, with a space-based mission in the form of EUCLID or WFIRST, microlensing could detect many Earth-mass habitable-zone planets and provide comprehensive statistics on the frequency of habitable bodies (often dubbed $\eta_{\oplus}$). On the down side, microlensing events are inherently transient and once the lensing event is over there are no further opportunities for follow-up.

Microlensing hit upon its first success in 2004 with the discovery of OGLE 2003-BLG-235/MOA 2003-BLG-53, named jointly as it was discovered by both the OGLE and MOA surveys. Bond et al. [39] reported the discovery of 1.5 M_J at 3 AU and since this time several more have been found.

1.3.8 Direct Imaging

The latest technique to broach success could also be considered the oldest technique used to detect a planet—direct imaging. Direct imaging seeks to either spatially, or spectrally, resolve the light from the planet from that of the star. The light from the planet is either reflected or thermal emission and so hot planets at a wide separation, meaning young planetary systems, are favourable. For reflected light only, the contrast ratio between the disk-averaged intensities of the planet and star is given by:

$$\begin{aligned}\frac{I_P}{I_*}\mathrm{d}\lambda &= \mathfrak{A}_G C(f_P)\frac{\pi R_P^2}{4\pi a_P^2}\mathrm{d}\lambda \\ &= 5.71\times 10^{-8}\left(\frac{R_P}{R_J}\right)\left(\frac{\mathrm{AU}}{a_P}\right)\mathfrak{A}_G C(f_P)\mathrm{d}\lambda\end{aligned} \tag{1.5}$$

Where $\mathfrak{A}_G$ is the geometric albedo, defined as the flux reflected by the planet when viewed at opposition (full phase), and $C(f_P)$ can be between 0 and 1 depending upon the true anomaly of the planet. Equation (1.5) therefore reveals that the contrast ratios are likely to be quite extreme.

Matters are further exacerbated by the challenge of resolving this light. The star and planet have an angular separation of a_P/d, where d is again the distance to the star from the observer. Typically, telescopes have their angular resolution constrained by the diffraction limit, given by $1.22\lambda/D$, where λ is the wavelength of the light one is using and D is the diameter of the telescope. Therefore, the smallest separation one can detect is:

$$\begin{aligned}a_{P,\min} &= \frac{1.22d\lambda}{D} \\ a_{P,\min} &= (1.38\mathrm{AU})\left(\frac{d}{10\ \mathrm{pc}}\right)\left(\frac{\lambda}{550\ \mathrm{nm}}\right)\left(\frac{1\ \mathrm{m}}{D}\right)\end{aligned} \tag{1.6}$$

So it can be seen that direct imaging naturally prefers to use large telescopes with shorter wavelengths to look around nearby stars with wide companions. Coronographic techniques can null the stellar flux leaving the planets alone to identify and track over the course of their orbit (e.g. [40]).

The first image of an extrasolar planet is 2M1207b by Chauvin at al. [41]. The primary is a brown dwarf meaning the contrast ratio between the planet and the "star" is quite favourable. Also, the planet orbits at ~50 AU around the primary and is hot, being in the range 1000–2000K. The pioneering first image is shown in Fig. 1.9.

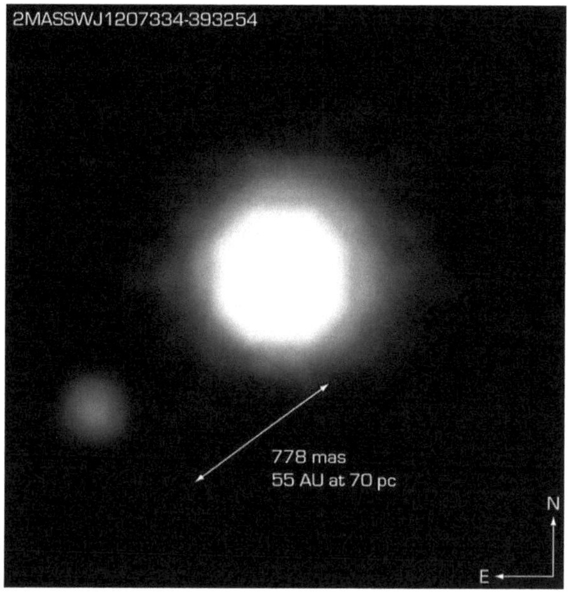

Fig. 1.9 First image of an extrasolar planet by Chauvin et al. [41]. 2M1207b orbits its primary, a brown dwarf named 2M1207a, at around 50 AU

1.4 Lessons from History

1.4.1 Context

This tour of the history of exoplanet discovery does not only serve to introduce this thesis but clearly exposes several key lessons which one can, and should, take away. In this thesis, I seek to develop ways to detect the moon of an extrasolar planet, which, it must be admitted, is a challenging task. There is little question that the signals being dealt with will exhibit very low signal-to-noise, just like the early years of exoplanet searches. In such a circumstance, how can one avoid the pitfalls of the early claims made for planets?

In this section, I ask, what are the differences between the discovery papers, methods and attitudes of those discoveries which were genuine and those which were not?

1.4.2 Detection Criteria of the Historically Genuine Discoveries

1.4.2.1 Statistical Significance

Looking back to the earliest claim of Jacob [10], there are many differences in style, presentation and evident resources applied to that of modern exoplanet detection papers. The first thing one can note is that the statistical significance of Jacob's claim

(1.9-σ) is marginal, at best, even assuming absolutely no systematic errors (which is of course never true). A very similar situation is true for Van de Kamp [18] whose model fit I have estimated to be accepted with 1.7-σ confidence.

In contrast, the successful first detections of Wolszczan and Frail [22], Mayor & Queloz [26] and Charbonneau et al. [34] exhibit significances of 13.3-, 8.9- and 31.8-σ respectively. Although detecting a signal to $\sim$10-σ confidence does not guarantee a genuine detection (e.g. the false positive of Bailes et al. [23]), it must certainly be considered as a requisite along the chain of steps necessary to claim a detection. I identify this as my first detection criterion and use 3-σ as an absolute minimum detection threshold.

1.4.2.2 Systematic Errors

Another clear difference is that Jacob [10] and Van de Kamp [18] give little, if any, consideration to the hypothesis that the observations are induced by some kind of systematic error(s).[7] In modern detection papers, this is a very regular practice. For example, RV detected planets are invariably accompanied with a bisector analysis to eliminate the hypothesis of a blended eclipsing binary (e.g. [26]). Frequently, they also come with a photometric campaign too (e.g. [42]) to confirm the signal is not due to star spots or plages (e.g. [43]). Similarly, planets detected through the transit method are usually given a thorough consideration of various blending scenarios and almost always come with RV confirmation (e.g. [44]).

I therefore identify a second criterion to be that an exploration of systematic effects, both instrumental and astrophysical, should be performed.

1.4.2.3 Physicality of the Claim

For a two-body system, almost all configurations can be considered plausible but once one moves into the realm of three-body systems then many configurations would quickly become dynamically unstable after a time span much shorter than the age of the systems in question. As an example of this, the claim of See [12] was challenged by Moulton [13] on the basis that the system would rapidly become unstable. Not surprisingly, observations later confirmed Moulton's warning [14]. Modern exoplanet papers regularly test the dynamical stability of the proposed system for multi-planet systems (e.g. [42]) as a criterion for the system being real. I take this as my third criterion.

1.4.2.4 Suspicious Periods

Bailes et al. [23] provide an interesting example of a signal cleanly detected to high significance but which ultimately proved to be false. The signal had a period of nearly

---

[7] Note that systematic errors can be either instrumental or astrophysical in origin.

exactly 6 months and turned out to be due to a calibration error for compensating for the Earth's orbit. Similarly, it can be seen to be prudent to check for signals which can be due to aliasing of the sampling frequency [45]. This gives the fourth criterion.

1.4.2.5 Consistent Instrumentation

The claim of Van de Kamp [18] was eventually shown to be false as a result of upgrades to the telescope over the course of the observations. These upgrades meant essentially a different instrument was being used between the start and end of the observations. If an instrument has an unknown systematic error which causes a constant offset then using it repeatedly actually has no bearing on the results, so long as the offset is indeed constant. However, if one uses different instruments, each one with its own unique (and unknown) systematic offset, then artificial signals will inevitably appear.

A lesson to take home is that it is strongly preferable to use the same instrument as much as possible for the observational data in a detection paper. This constitutes the fifth criterion.

1.4.2.6 Avoid Observing Targets with Large Systematic Offsets in the Observed Parameter

The false claim of Van de Kamp [18] resonates with a later claim made for a different star, VB 10, in 2009. Pravdo and Shaklan [46] claimed to have detected a planet around VB 10 using astrometry, in what would have been the first confirmed detection using this method. However, it was later shown by Anglada-Escudé et al. [47] and Bean et al. [48] that the predicted RV signal did not exist and thus the astrometric signal was spurious.

What VB 10 and Barnard's star have in common is that both are objects with very high proper motion (1" and 10.3" per year respectively). To perform astrometry, one therefore has to first remove this large systematic effect of the proper motion. One can easily see that if there is a, say 0.01%, error in this process then a residual signal of 0.1 mas will persist in the case of Barnard's star, which is the same size as the claimed planetary signal for that system.

Therefore, targeting objects with extremely high systematics (i.e. proper motion in the case of astrometry) can be instantly appreciated to be an unwise strategy since it requires exceptionally accurate corrections for these effects before a meaningful analysis can be performed. This constitutes the sixth criterion.

1.4.3 Summary

I have identified six criteria which seem to discriminate between the genuine and spurious exoplanet claims over the past decades. I caution that this is by no means comprehensive—counter examples of genuine systems will surely exist for any given criterion—but they do offer a general strategy to optimize the chances of finding authentic discoveries by exploiting the benefit of hindsight.

C1 Statistical significance
C2 Systematic errors
C3 Physicality of the claim
C4 Suspicious periods
C5 Consistent instrumentation
C6 Avoid large systematics

1.4.4 From Exoplanets to Exomoons

To date, over 500 extrasolar planets have been detected.[8] The lower mass limits continue to drop with Super-Earth detections becoming routine and Earth-mass objects appearing imminent. As we approach the Earth-mass and radius limit, it is natural to consider the possibility of detecting the moons of exoplanets too.

Satellite systems represent mini Solar Systems, with a richness and diversity all of their own. The exploration of the moons of Jupiter and Saturn has reaped immense understanding and shown us worlds beyond our imagination. The prospect of studying thousands of new examples in utterly alien environments would provide unforeseeable understanding of our place in the Universe.

The problem I therefore identify in this thesis, is how can we make this ambitious vision a reality? How can we detect and characterize the moons of extrasolar planets?

1.4.5 Thesis Outline

In this thesis, I aim to provide a method for detecting the moons of transiting extrasolar planets. To achieve this goal it will be necessary to first understand the transit light curve of an exoplanet without a moon and so this thesis will be broadly split into two objectives i) an exploration of methods to model the transit light curve ii) description of a novel method to detect the moons of transiting planets.

In Chap. 2, I will discuss the motivations for such a search, the expected properties of an exomoon and possible methods which could be used to detect the moons of non-transiting planets. Having established what is being searched for and why, I move on to describe the basics of transit theory in Chap. 3, which must be first understood. The method to detect moons will be later be shown to rely on precise timing of transits and so in Chap. 4 I derive the times of transit minima and the transit durations. Chapter 4 marks the start of my own research into this field with the other material covered thus far being predominantly introductory. In Chap. 5, this continues with an exploration of two subtleties with modelling the transit light curve. I discuss how hot planets can produce nightside emission which acts as a self-blend, which acts to dilute the transit

[8] See http://www.exoplanet.eu

depth. Then, I explore how long-integration times, such as that used by *Kepler*'s long-cadence mode cause a smearing of the transit signal and provide methods to compensate for the distortion.

With the theory of transiting planets well established by the end of Chap. 5, the stage is set to introduce the exomoon into the model. In Chap. 6, I present the transit timing variations (TTV) method of detecting exomoons as first proposed by Sartoretti and Schneider [49]. I then go on to discuss two critical problems with TTV, namely TTV only provides the mass of the moon multiplied by the orbital distance, $M_S a_S$, and also how many other phenomena can also cause TTVs leading to a signal ambiguity. I then predict the existence of a new timing effect, velocity induced transit duration variations (TDV-V), which solves both of these problems. TDV-V scales as $M_S a_S^{-1/2}$ and thus the ratio of TDV-V to TTV solves for M_S and a_S separately. Further, TDV-V exhibits a $\pi/2$ phase shift from TTV and thus provides a unique exomoon signature. The chapter closes by exploring a second order TDV effect, dubbed transit impact parameter induced transit duration variations (TDV-TIP), which could be exploited to determine an exomoon's sense of orbital motion with sufficient signal-to-noise.

Whilst the theory of transit timing effects is presented in Chap. 6, the issue of how feasible an exomoon detection would be with these techniques remains absent. To address this, Chap. 7 provides a feasibility study of detecting habitable-zone exomoons with *Kepler*-class photometry. In this chapter, I conclude that exomoons down to 0.2 $M_{\oplus}$ could be detected in the most favourable scenarios and up to 25,000 stars within *Kepler's* field-of-view are bright enough to be surveyed for 1 $M_{\oplus}$ habitable-zone exomoons. These encouraging figures mean that TTV and TDV have been shown to be viable detection techniques for finding exomoons.

I conclude in Chap. 8 with a summary of the thesis and a look at future work required in exolunar theory. Notably, a work in preparation, which can model the transit dips of exomoons, is discussed and the potential for it to characterize exomoons.

References

1. C. Ptolemy, Almagest, 1st edn. (147)
2. W. Herschel, Account of a Comet, Mr. Herschel, F. R. S.; Communicated by Dr. Watson, Jun. of Bath, F. R. S. Phil. Trans. Roy. Soc. London, **71**, 492–501 (1781)
3. U. Le Verrier, M. Lettre de, M. Le Verrierá, Faye sur la théorie de Mercure et sur le mouvement du périhélie de cette planéte. Comptes rendus hebdomadaires des séances de l'Académie des sciences **49**, 379–383 (1859)
4. J.G. Galle, Account of the discovery of Le Verrier's planet Neptune, Berlin, Sept. 23, 1846. MNRAS, **7**, 153–153 (1846)
5. I. Newton, Philosophi Naturalis Principia Mathematica, 1st edn. (1687)
6. G.M. Clemence, The relativity effect in planetary motions. Rev. Mod. Phys. **19**, 361–364 (1947)
7. A. Einstein, Die Feldgleichungen der Gravitation. Sitzungsberichte der Königlich Preußischen Akademie der Wissenschaften (Berlin) Seite 844–847 (1915)
8. A. Bouvard, Tables astronomiques publices par le Bureau des Longitudes de France. 1st edn. (Bachelier, Paris, 1821)

9. J.C. Adams, Explanation of the observed irregularities in the motion of Uranus, on the hypothesis of disturbance by a more distant planet. MNRAS **6**, 149–149 (1846)
10. W.S. Jacob, On certain Anomalies presented by the Binary Star 70 Ophiuchi. MNRAS **15**, 228–230 (1855)
11. K. Pearson, On the criterion that a given system of deviations from the probable in the case of a correlated system of variables is such that it can be reasonably supposed to have arisen from random sampling. Phil. Mag. **50**, 157–175 (1900)
12. T.J.J. See, Researches on the orbit of 70 Ophiuchi, and on a periodic perturbation in the motion of the system arising from the action of an unseen body. AJ **16**, 17–23 (1896)
13. F.R. Moulton, The limits of temporary stability of satellite motion, with an application to the question of the existence of an unseen body in the binary system. AJ **20**, 33–36 (1899)
14. W.D. Heintz, The binary star 70 Ophiuchi revisited. JRASC **82**, 140–145 (1988)
15. C.D. Murray, S.F. Dermott, Solar system dynamics. (Cambridge University Press, Cambridge, 1999)
16. F. Bessel, Bestimmung der Enterfernung des 61 Sterns des Schwans. Astronomische Nachrichten **1**, 365–381 (1838)
17. F. Bessel, On the proper motions of Procyon and Sirius. MNRAS **6**, 136–136 (1844)
18. P. Van de Kamp, Astrometric study of Barnard's star from plates taken with the 24-inch Sproul refractor. AJ **68**, 515–521 (1963)
19. G. Gatewood, H. Eichhorn, An unsuccessful search for a planetary companion of Barnard's star BD +4 3561. AJ **78**, 769–776 (1973)
20. J.L. Hershey, Astrometric analysis of the field of AC +65 6955 from plates taken with the Sproul 24-inch refractor. AJ **78**, 421–425 (1973)
21. G.F. Benedict, B.E. McArthur, T. Forveille, X. Delfosse, E. Nelan, R.P. Butler, W. Spiesman, G. Marcy, B. Goldman, C. Perrier, W.H. Jefferys, M. Mayor, A mass for the extrasolar planet Gliese 876b determined from Hubble space telescope fine guidance sensor 3 astrometry and high-precision radial velocities. ApJL **581**, L115–L118 (2002). doi:10.1086/346073
22. A. Wolszczan, D.A. Frail, A planetary system around the millisecond pulsar PSR1257+12. Nature **355**, 145–147 (1992)
23. M. Bailes, A.G. Lyne, S.L. Shemar, A planet orbiting the neutron star PSR1829 − −10. Nature **352**, 311–313 (1991)
24. M. Tuomi, S. Kotiranta, M. Kaasalainen, The complementarity of astrometric and radial velocity exoplanet observations. Determining exoplanet mass with astrometric snapshots. A&A **494**, 769–774 (2009)
25. J.B. Pollack, O. Hubickyj, P. Bodenheimer, J.J. Lissauer, M. Podolak, Y. Greenzweig, Formation of the giant planets by concurrent accretion of solids and gas. Icarus **124**, 62–85 (1996)
26. M. Mayor, D. Queloz, A Jupiter-mass companion to a solar-type star. Nature **378**, 355–359 (1995)
27. B. Campbell, G.A.H. Walker, S. Yang, A search for substellar companions to solar-type stars. ApJ **331**, 902–921 (1988). doi:10.1086/166608
28. G.A.H. Walker, D.A. Bohlender, A.R. Walker, A.W. Irwin, S.L.S. Yang, A. Larson, Gamma Cephei—Rotation or planetary companion? ApJL **396**, 91–94 (1992). doi:10.1086/186524
29. A.P. Hatzes, W.D. Cochran, M. Endl, B. McArthur, G.A.H. Paulson, D.B. Walker, B. Campbell, S. Yang, A planetary companion to γ Cephei A. ApJ **599**, 1383–1394 (2003). doi:10.1086/379281
30. D.W. Latham, R.P. Stefanik, T. Mazeh, M. Mayor, G. Burki, The unseen companion of HD114762—A probable brown dwarf. Nature **339**, 38–40 (1989). doi:10.1038/339038a0
31. G.W. Marcy, R.P. Butler, A planetary companion to 70 Virginis. ApJL **464**, L147–L147 (1996)
32. R.P. Butler, G.W. Marcy, A planet orbiting 47 Ursae Majoris. ApJL **464**, L153–L153 (1996)
33. S.S. Vogt, R.P. Butler, E.J. Rivera, N. Haghighipour, G.W. Henry, M.H. Williamson, *The Lick-Carnegie exoplanet survey: A 3.1 M_Earth Planet in the Habitable Zone of the Nearby M3V Star Gliese 581* (ArXiv e-prints, 2010a)

34. D. Charbonneau, T.M. Brown, D.W. Latham, M. Mayor, Detection of planetary transits across a Sun-like star. ApJL **529**, L45–L48 (2000)
35. G.W. Henry, G.W. Marcy, R.P. Butler, S.S. Vogt, A transiting "51 Peg-like" planet. ApJL **529**, L41–L44 (2000)
36. G. Bakos, R.W. Noyes, G. Kovács, K.Z. Stanek, D.D. Sasselov, I. Domsa, Wide-field millimagnitude photometry with the HAT: a tool for extrasolar planet detection. PASP **116**, 266–277 (2004)
37. D.L. Pollacco, I. Skillen, A.C. Cameron, D.J. Christian, C. Hellier, J. Irwin, T.A. Lister, R.A. Street, R.G. West, D. Anderson, W.I. Clarkson, H. Deeg, B. Enoch, A. Evans, A. Fitzsimmons, C.A. Haswell, S. Hodgkin, K. Horne, S.R. Kane, F.P. Keenan, P.F.L. Maxted, A.J. Norton, J. Osborne, N.R. Parley, R.S.I. Ryans, B. Smalley, P.J. Wheatley, D.M. Wilson, The WASP project and the superWASP cameras. PASP **118**, 1407–1418 (2006)
38. T.M. Brown, Transmission spectra as diagnostics of extrasolar giant planet atmospheres. ApJ **553**, 1006–1026 (2001)
39. I.A. Bond, A. Udalski, M. Jaroszyński, N.J. Rattenbury, B. Paczyński, I. Soszyński, L. Wyrzykowski, M.K. Szymański, M. Kubiak, O. Szewczyk, K. Żebruń, G. Pietrzyński, F. Abe, D.P. Bennett, S. Eguchi, Y. Furuta, J.B. Hearnshaw, K. Kamiya, P.M. Kilmartin, Y. Kurata, K. Masuda, Y. Matsubara, Y. Muraki, S. Noda, K. Okajima, T. Sako, T. Sekiguchi, D.J. Sullivan, T. Sumi, P.J. Tristram, T. Yanagisawa, P.C.M. Yock, OGLE 2003-BLG-235/MOA 2003-BLG-53: A Planetary Microlensing Event. ApJL **606**, L155–L158 (2004)
40. S.R. Heap, D.J. Lindler, T.M. Lanz, R.H. Cornett, I. Hubeny, S.P. Maran, B. Woodgate, Space telescope imaging spectrograph coronagraphic observations of β Pictoris. ApJ **539**, 435–444 (2000)
41. G. Chauvin, A.-M. Lagrange, C. Dumas, B. Zuckerman, D. Mouillet, I. Song, J.-L. Beuzit, P. Lowrance, A giant planet candidate near a young brown dwarf. Direct VLT/NACO observations using IR wavefront sensing. A&A **425**, L29–L32 (2004)
42. S.S. Vogt, R.A. Wittenmyer, R.P. Butler, S. O'Toole, G.W. Henry, E.J. Rivera, S. Meschiari, G. Laughlin, C.G. Tinney, H.R.A. Jones, J. Bailey, B.D. Carter, K. Batygin, A Super-Earth and two Neptunes orbiting the nearby Sun-like star 61 virginis. ApJ **708**, 1366–1375 (2010)
43. D. Queloz, G.W. Henry, J.P. Sivan, S.L. Baliunas, J.L. Beuzit, R.A. Donahue, M. Mayor, D. Naef, C. Perrier, S. Udry, No planet for HD 166435. A&A **379**, 279–287 (2001)
44. G.Á. Bakos, G. Torres, A. Pál, J. Hartman, G. Kovács, R.W. Noyes, D.W. Latham, D.D. Sasselov, B. Sipőcz, G.A. Esquerdo, D.A. Fischer, J.A. Johnson, G.W. Marcy, R.P. Butler, H. Isaacson, A. Howard, S. Vogt, G. Kovács, J. Fernandez, A. Moór, R.P. Stefanik, J. Lázár, I. Papp, P. Sári, HAT-P-11b: a Super-Neptune planet transiting a bright K star in the kepler field. ApJ **710**, 1724–1745 (2010)
45. R.I. Dawson, D.C. Fabrycky, Radial velocity planets De-aliased: a new, short period for Super-Earth 55 Cnc e. ApJ **722**, 937–953 (2010)
46. S.H. Pravdo, S.B. Shaklan, An ultracool star's candidate planet. ApJ **700**, 623–632 (2009)
47. G. Anglada-Escudé, E.L. Shkolnik, A.J. Weinberger, I.B. Thompson, D.J. Osip, J.H. Debes, Strong constraints to the putative planet candidate around VB 10 using Doppler spectroscopy. ApJL **711**, L24–L29 (2010)
48. J.L. Bean, A. Seifahrt, H. Hartman, H. Nilsson, A. Reiners, S. Dreizler, T.J. Henry, G. Wiedemann, The proposed giant planet orbiting VB 10 does not exist. ApJL **711**, L19–L23 (2010)
49. P. Sartoretti, J. Schneider, On the detection of satellites of extrasolar planets with the method of transits. A&AS **134**, 553–560 (1999)
50. A. Hewish, S.J. Bell, J.D.H. Pilkington, P.F. Scott, R.A. Collins, Observation of a rapidly pulsating radio source. Nature, **217**, 709–713, (1968). doi:10.1038/217709a0, http://adsabs.harvard.edu/abs/1968Natur.217..709H (Provided by the SAO/NASA Astrophysics Data System)

Chapter 2
Extrasolar Moons

I think we're going to the Moon because it's in the nature of the human being to face challenges

Neil Armstrong, Apollo mission press conference, 1969

2.1 Motivations

Extrasolar moons, or "exomoons", offer a significant challenge to astronomers given their anticipated[1] low masses and radii. However, there are many motivations for looking for such objects, which I will overview here.

2.1.1 *Intrinsic Habitability*

Based upon the only known examples, moons are likely to have either a solid or liquid surface (see Sect. 2.2 for a more detailed discussion of expected compositions). For that reason, exomoons have the same capacity as Earths and Super-Earths (planets of mass $\lesssim 10\ M_{\oplus}$) for harbouring a habitable environment.

Habitability of exomoons is discussed in more detail in [1], who find that such bodies can offer suitable conditions provided they are massive enough to maintain an atmosphere for Gyr. The authors argue that this cut-off occurs approximately at $\gtrsim 0.3\ M_{\oplus}$.

Scharf [2] explored the interesting possibility of moons outside of the conventional habitable-zone being tidally heated to an extent where they could maintain oceans of liquid water, which may be suitable for life.

[1] From a Copernican perspective

D. M. Kipping, *The Transits of Extrasolar Planets with Moons*, Springer Theses,
DOI: 10.1007/978-3-642-22269-6_2,

2.1.2 *Extrinsic Habitability*

Even if a moon is not habitable itself, a moon could enhance the probability of a host planet maintaining a habitable environment. As an example, Laskar and Robutel [3] have argued that the presence of a large satellite around the Earth (i.e. the Moon) stabilizes the axial tilt of our planet and thus leads to a more favourable long-term climate for life. This may not significantly affect the probability of the emergence of life, but is thought to affect the probability of land-based life emerging [4].

Benn [4] provide a review of the various proposed beneficial influences the Moon may have had on the emergence of life on the Earth:

- Stabilization of the Earth's axial tilt [3]
- Elimination of the primordial atmosphere [5, 6]
- Generation of the Earth's magnetic field [7]
- Generation of large tides [8]
- Generation of longer-period tides [9]

2.1.3 *Planet/Moon Formation*

The formation theory of the planets and their moons is a topic of active debate. Currently, two hypotheses of planet formation are considered the most viable (i) core accretion [10] (ii) gravitational instability [11].

Two mechanisms (although quite distinct from the planet formation models) have also been proposed for the formation of satellites. Moons are predicted to form either from the disk material surrounding a planet ("regular satellites"; always prograde) [12] or by gravitational capture/impacts/exchange interactions ("irregular satellites"; either prograde or retrograde) [13].

The detection of many extrasolar examples of satellite systems would undoubtedly enhance our understanding of the formation of planets and moons by providing a greater population sample. This is particularly salient for the large irregular satellites, as currently only two examples are known to exist, the Moon and Triton. Whilst Triton is likely to have been gravitationally captured by Neptune [14], the most plausible mechanism for the formation of the Moon is quite distinct. In this case, a Mars-size planet, dubbed "Theia", is thought to have impacted the primordial Earth [6] (see Fig. 2.1) and the prevalence of such encounters would impinge powerfully on models of young planetary systems.

2.2 Predicted Properties of Exomoons

Having established a clear motivation for searching for exomoons (Sect. 2.1), one may begin to consider how they might be found. However, it is useful to first overview the expected properties of exomoons, which will inform the choice of detection method.

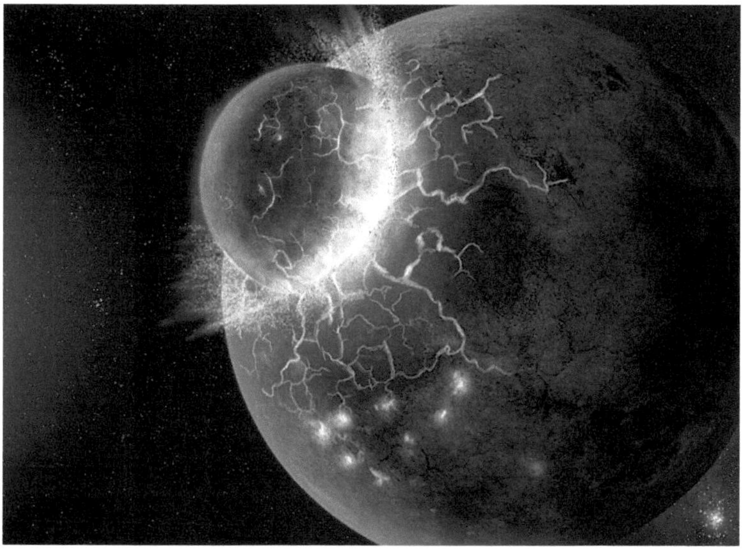

Fig. 2.1 Artist's impression of the impact of a Mars sized planet (dubbed "Theia") into the primordial Earth. This "Big Splash" scenario is currently believed to be the most plausible mechanism for the formation of the Moon. Image credit Fahad Sulehria (http://www.novacelestia.com)

2.2.1 Formation

As discussed in Sect. 2.1.3, current models of the formation of satellites is naturally based upon observations of the Solar System. There appear to be two broad classes of moons; regular and irregular satellites. Canup [15] have proposed that the maximum total mass of a satellite system, formed as regular satellites, is $\sim 2 \times 10^{-4}$ times the mass of the host planet. This is based upon the expected quantity of material in the disc surrounding a planet early in its history, which may subsequently be gravitationally bound to the planet and coalesce into solid moons. The region in which a massive body, in orbit of a star, dominates the attraction of surrounding test particles is known as the Hill sphere (named after George William Hill who provided this definition). The radius of this sphere is found by solving the restricted circular three-body problem and is equal to the distance of the L1/L2 Lagrange points:

$$R_H = a_P \left(\frac{M_P}{3M_*} \right)^{1/3} \tag{2.1}$$

Given that the focus of this thesis is detecting moons, and it is known that even detecting a Super-Earth mass planet is challenging with current instrumentation, one must acknowledge that the most favourable, and in all likelihood even feasible, case will be a single moon around a planet. This is because the total allowed mass will be concentrated into a single body. The Saturn–Titan system is perhaps the closest example of this whereas Jupiter has four large moons with roughly equivalent masses.

Following the hypothesis of [15], to reach an Earth-mass Moon, one would therefore need a host planet of mass $\sim 16\ M_J$, which is really a brown dwarf rather than a "planet".

The second formation mechanism is that of the irregular satellites. Classifying an object as such does not really constrain the expected properties in many ways. The mass of such an object could be anything from an asteroid to a Super-Earth.[2] The latter clearly would require a full-grown planet to get captured by a gas giant planet. Similarly, this classification does not strongly constrain the origins of the satellite, which could be anything from an impact scenario (e.g. Earth-Moon [6]) to a capture (e.g. Triton-Neptune [14]).

This vast range of possible histories actually means that irregular satellites are quite favourable for an exomoon search. The permissible existence of Earth mass moons around "normal" gas giants (i.e. not brown dwarfs) certainly increases the chances of a detection. By this reasoning, I consider it much more likely that the first detected exomoon will be an irregular satellite, based upon our current understanding of the regular satellites.

I also mention that the existence of binary-planets is physically sound. Podsiadlowski [16] showed that one viable scattering history in the formation of a planetary system is the tidal capture of two planets forming a binary. Indeed, a Jupiter–Earth pair could be considered as an extreme binary, much like Pluto-Charon.

2.2.2 Evolution

Having identified that irregular satellites are the most probable bodies which can be detected in the near-future, let us consider the chances of a moon's survival. Two dominant effects are believed to cause the loss of moons for a single planet–moon pair[3]: (i) tidal dissipation (ii) inward planetary migration.

Tidal dissipation between a planet and its companion cause a moon to either spiral in or out. Which of these occurs depends on whether the tidal bulge caused by the moon leads (spirals-in) or lags (spirals-out) relative to the rotation bulge of the planet. Barnes and O'Brien [17] provide analytic approximations for the maximum time a moon can survive and encouragingly predict that Earth-mass moons are stable around Jupiter-like planets for habitable-zone periods for Sun-like stars. Planets on short periods have tighter stable regions for a moon to reside and thus the planet–moon tides are greater leading to a rapid loss i.e. hot-Jupiters are unfavourable.

The other effect is that of inward, disk-driven planetary migration, which occurs on a much faster time scale than tidal dissipation. Because of this fact, the Hill radius of the planet shrinks very quickly. Thus, a moon initially well inside the Hill sphere can find itself outside and so be ejected. Namouni [18] showed that once a migrating

---

[2] Although, of course, no examples of this exist in the Solar System.

[3] Multiple moons are more complex due to the dynamical interactions which occur between the moons.

gas giant crosses $\sim$0.1 AU, a moon is unlikely to survive. For this reason, hot-Jupiters are again unfavourable hosts.

2.2.3 Stable Region

I earlier provided the distance at which a test particle feels a greater gravitational attraction to the planet rather than the star- the Hill sphere (see Sect. 2.2.1 and Eq. 2.1). However, this simple picture does not accurately represent the complete stable region of an extrasolar moon. Other effects, such as the Yarkovsky effect [19] and three-body perturbations can eventually perturb a moon to be ejected from the system.

Domingos et al. [20] showed that the actual region over which a moon is stable depends upon the eccentricity of the orbit and the sense of orbital motion. For prograde and retrograde moons respectively, the stable regions are:

$$a_{S,\max}^{\text{prograde}} = 0.4895 R_H (1.0000 - 1.0305 e_P - 0.2738 e_S) \tag{2.2}$$

$$a_{S,\max}^{\text{retrograde}} = 0.9309 R_H (1.0000 - 1.0764 e_P - 0.9812 e_S) \tag{2.3}$$

where e_P and e_S are the orbital eccentricities of the planet and satellite respectively. Therefore, retrograde moons can be found at far larger distances than their prograde counterparts. This means the moon has a larger distance over which to tidally spin-out or in and thus ultimately a longer lifetime. This is equivalent to allowing a larger moon mass to be stable for a given amount of time. One can also see that planets on eccentric orbits offer severely reduced regions of stability for a putative moon.

More recently, Donnison [21] showed that moons on inclined orbits also yield contracted regions of orbital stability and thus in general one expects moons to be roughly co-aligned to the orbital plane of the planetary orbit, which is supported by the properties of the massive satellites in our own solar system.

2.2.4 Composition

For the regular satellites of the Solar System, the composition of a moon tends to be ice-rich with the rest being silicates and iron [22]. This is because gas giants are thought to have formed beyond the snow-line, where ices do not sublimate away and thus the moons which form in situ (i.e. regular moons) have plenty of ice to accumulate in their formation [10]. However, given that this thesis has identified the best chance of detecting an exomoon to be a terrestrial planet which becomes captured around a gas giant, the composition of a detected exomoon is more likely to resemble Mercury, Venus, Earth and Mars. Useful mass-radius scalings for such bodies are presented in [23], who estimate $(R/R_{\oplus}) \sim (M/M_{\oplus})^{0.27}$.

2.2.5 Conclusions

The largest and most massive (and thus most detectable) possible satellite which could exist is likely to be a terrestrial planet which was captured as an irregular moon around a gas giant planet. This statement merely indicates physically feasible origin for an exomoon which could be detected in the coming years. The consequences are that the moon is likely to have a similar composition and mass scaling as the terrestrial planets of the Solar System and may exist in a retrograde orbit (e.g. Triton). Further, the host planet is likely to be on an orbit with semi-major axis $\gtrsim$0.1 AU with low eccentricity.

2.3 Proposed Detection Methods for Non-Transiting Systems

In this thesis, I will present methods of detecting moons using transit light curves. However, other methods are of course possible for non-transiting systems as well. In this section, I will discuss the different methods available, which will provide a context for the method of transits.

2.3.1 Direct Imaging

Direct imaging of an exomoon would face similar, but ultimately amplified, problems as imaging an exoplanet. For the latter, one needs to spatially resolve the star and planet which have an angular separation of a_P/d. However, for a moon one needs to acquire an angular resolution of a_S/d. Given that the maximum allowed $a_S \simeq R_H$ (i.e. the Hill radius), then a simple calculation reveals the angular resolution requirement is at least a factor of $(3M_*/M_P)^{1/3}$ lower. For a Jupiter around a Sun-like star this is a factor of 14.4 i.e. comfortably an order-of-magnitude more challenging.

The glare of the host star also becomes more problematic as a 1 $R_\oplus$ exomoon would have a contrast ratio of $\sim$4.5 $\times$ 10^{-10} to the host star (using Eq. 1.5 and assuming a_P = 1 AU). This is a factor of 125 times more difficult than a Jovian planet i.e. two orders of magnitude.

2.3.2 Microlensing

Microlensing does not offer the ability to characterize moons once the lensing event is over and thus is more useful for understanding the statistical distributions of moons rather than their nature.

Liebig and Wambsganss [24] recently calculated the detectability of an Earth mass moon around a Saturn-like planet at 2.5 AU for various configurations using the microlensing technique. The authors find that for fairly rapid sampling of once every 15 minutes with an RMS photometric precision of 20 mmag, there is a 30% chance of detecting such a system. As the planet-star separation is decreased, this probability rapidly falls and the authors conclude only "cold" planets are feasible for an exomoon detection in the near future.

Nevertheless, the case for a microlensing detection of an exomoon is strong when considered in light of the planned space missions (EUCLID/WFIRST). However, due to the fact that a habitable-zone exomoon is unlikely to be found and that follow-up characterization is not possible, the microlensing prospect also has a lot to be desired.

2.3.3 Pulsar Timing

Is there any way that studying the motion of the host star could reveal the presence of a moon? At the simplest level, the reflex motion of a star is exactly the same for a 1 M_J planet with a 0.003 M_J moon as it is for a 1.003 M_J planet without a moon. In other words, the star sees the planet-moon system as a composite point-mass and thus there is no way one could detect a moon by watching the motion of the host star.

However, this is not strictly true and the presence of the moon does induce slight deviations in the star's reflex motion. These tiny deviations arise from a full analysis of the three body problem. In reality, the moon and planet are not a single composite point-mass but two distinct masses which give rise to a varying torque on the host star.

Lewis et al. [25] first considered the possibility of detecting moons in this way, for pulsars with a time-of-arrival (TOA) analysis. The planet plus moon signal has two components (i) a dominant term which essentially acts like $M_P + M_S$ in orbit of the host star and thus indistinguishable from a slightly more massive planet (ii) a residual term which may be used to detect the moon. This residual signal is given by Lewis et al. [25]:

$$\begin{aligned} \mathrm{TOA}_{\mathrm{residual}} = &- \sin i_P \frac{M_P M_S}{(M_P + M_S)^2} \frac{a_P}{c} \left(\frac{2a_S}{a_P}\right)^5 \\ &\times \left[\frac{3}{32}\cos(\phi_P - 2\phi_S) + \frac{15}{32}\cos(3\phi_P - 2\phi_S)\right] \end{aligned} \tag{2.4}$$

where ϕ_P is the angular position of the planet-moon barycentre around the host star and ϕ_S is the angular position of the satellite around the planet-moon barycentre. Thus, the residual signal has two frequency components at $(n_P - 2n_S)$ and $(3n_P - 2n_S)$, where n_P denotes the mean motion of the planet-moon barycentre around the star and n_S denotes the mean motion of the satellite around the planet-moon barycentre. Replacing a_S with the Hill radius, and assuming $M_P \gg M_S$, the maximized amplitude of this signal can be expressed as:

$$
\begin{aligned}
|\mathrm{TOA}_{\mathrm{residual}}| &= \frac{9 \sin i_P}{16} \frac{M_P M_S}{(M_P + M_S)^2} \frac{a_P}{c} \left(\frac{2 a_S}{a_P} \right)^5 \\
&\simeq \frac{9 \sin i_P}{16} M_S \frac{a_P}{c} \frac{M_P^{2/3}}{3^{5/3} M_*^{5/3}} \\
&= (41918 \mathrm{ns}) \sin i_P \left(\frac{M_S}{M_\oplus} \right) \left(\frac{a_P}{\mathrm{AU}} \right) \left(\frac{M_P}{M_J} \right)^{2/3} \left(\frac{M_\odot}{M_*} \right)^{5/3}
\end{aligned} \tag{2.5}
$$

This may be compared to the residuals of one of the most stable pulsars, PSR J0437-4715, which has an RMS of 130 ns [26]. Therefore, the detection of an Earth-mass moon would seem highly feasible. However, the ratio of the amplitude of this residual signal to the planet-only signal is $\simeq 0.347 M_P^{1/3} M_*^{2/3} M_S^{-1}$, or a factor of 11400 for a 1 M_J planet at 1 AU with an Earth mass moon. This ratio perhaps gives a more realistic quantification of the challenge faced.

Another drawback with the proposed method is that only two examples of pulsars with planets are known to exist, PSR B1257+12 and PSR B1620-26 and therefore one does not have many planets to investigate. The first of these is a multi-planet system where the approximations of [25] are invalid (i.e. a single planet with a single moon) [27]. The second is actually a binary star system with a poorly characterized planetary companion ($M_P = 2.5 \pm 1.0 M_J$) [28]. Thus, success does not seem likely to be forthcoming for the prospect of pulsar moons.

2.3.4 Astrometry

I briefly point out that the TOA signal from pulsar timing is an astrometric effect and thus a very similar equation exists for the astrometric amplitude. One can show that the astrometric signal is:

$$
\begin{aligned}
|\alpha_{\mathrm{residual}}| &= \frac{c |\mathrm{TOA}_{\mathrm{residual}}|}{d} \\
&= \frac{9 \sin i_P}{16} M_S \frac{a_P}{d} \frac{M_P^{2/3}}{3^{5/3} M_*^{5/3}} \\
&= (0.147 \mathrm{nas}) \left(\frac{d}{10 \mathrm{pc}} \right) \left(\frac{M_S}{M_\oplus} \right) \left(\frac{a_P}{\mathrm{AU}} \right) \left(\frac{M_P}{M_J} \right)^{2/3} \left(\frac{M_\odot}{M_*} \right)^{5/3}
\end{aligned} \tag{2.6}
$$

where "nas" indicates nano-arcseconds. Whilst a dedicated astrometric space mission, such as SIM, may be launched in the next 10–20 years, a sensitivity of at best μas would be achievable with such an enterprise [29]. Consequently, I conclude that achieving the nano-arcsecond precision needed to locate a moon is unfeasible in the near future.

2.3.5 Radial Velocity

In a similar manner to the astrometric signature predicted by Lewis et al. [25], the torque supplied by a binary should induce a residual radial velocity signature over the bulk centre-of-mass signal. This torque was investigated by Moris and Correia [30], who performed numerical simulations to compute the predicted RV signals for a variety of possible configurations.

The authors found that a Jupiter-Jupiter binary at 1 AU would induce a residual RV signal of 10^{-5} to 10^{-6} m/s and a Jupiter-Earth system is even more challenging at 10^{-7} to 10^{-8} m/s. These magnitudes are certainly far below the typical intrinsic noise level of even the quietest stars ($\sim$1 m/s, [31]), the so-called stellar "jitter" which is due to spots, plages and magnetic activity. In conclusion, a RV detection appears quite unfeasible in the near-future.

2.4 Summary

Amongst the myriad of motivations for embarking on a search for exomoons, I have identified three examples which outline the importance of this work. Section 2.1 discusses the potential of moons to be intrinsically habitable (Sect. 2.1.1) as well as increasing the probability of complex life prospering on the host planet (Sect. 2.1.2). Away from astrobiology, moons would also shed new light onto the theories of planet/moon formation (Sect. 2.1.3)

The expected properties of exomoons is discussed in Sect. 2.2, including their formation (Sect. 2.2.1), evolution (Sect. 2.2.2), dynamical stability (Sect. 2.2.3) and expected composition (Sect. 2.2.4). I conclude the most feasible and detectable case is a terrestrial planet which is captured by a Jupiter as an irregular satellite. This therefore means retrograde orbits are possible for the moon and it would likely exhibit a composition similar to the rocky planets of our own Solar System.

Finally, I have explored the possible methods of detecting a non-transiting planet. Direct imaging (Sect. 2.3.1), astrometry (Sect. 2.3.4) and radial velocity methods (Sect. 2.3.5) can be seen to be unfeasible as a method for detecting exomoons in the near future. Pulsar timing (Sect. 2.3.3) has the necessary sensitivity but an insufficient catalogue of planets to survey.

Microlensing (Sect. 2.3.2) has the greatest chance of detecting an exomoon in a non-transiting system in the near future. This seems to require space-based photometry to give a reasonable chance of success and indeed a mission such as EUCLID/WFIRST [32] looks likely to fly within a decade. However, microlensing is unlikely to identify habitable-zone exomoons in the near-future and would not be able to offer opportunities for intensive follow-up (e.g. to look for bio-signatures).

There is clearly a strong desideratum for a method which can detect exomoons, which have not only the potential to be habitable, but also be subsequently investigated to verify/reject such a hypothesis. Transiting systems, then, may offer the only hope of a near-future achievement of such a goal. However, before one can even entertain the notion of looking for exomoons with the transit method, one must first properly understand the transit light curve of a single planet, which I will discuss in the next chapter.

References

1. D.M. Williams, J.F. Kasting, R.A. Wade, Habitable moons around extrasolar giant planets. Nature **385**, 234–236 (1997)
2. C.A. Scharf, The potential for tidally heated icy and temperate Moons around exoplanets. ApJ **648**, 1196–1205 (2006)
3. J. Laskar, P. Robutel, The chaotic obliquity of the planets. Nature **361**, 608–612 (1993)
4. C.R. Benn, The Moon and the origin of life. EM&P **85**, 61–66 (2001)
5. A.G.W. Cameron, W. Benz, The origin of the moon and the single impact hypothesis-IV. Icarus **92**, 204–216 (1991)
6. S.R. Taylor, Solar system evolution: a new perspective. An inquiry into the chemical composition, origin, and evolution of the solar system. (Cambridge University Press, NewYork, 1992)
7. J. Pearson, New Scientist **2508**, 38–40 (1988)
8. C.F. Chyba, Impact delivery and erosion of planetary oceans in the early inner solar system. Nature **343**, 129–133 (1900)
9. R.T. Rood, J.S. Trefil, *Are We Alone: The Possibility of Extraterrestrial Eivilizations*. (Scribner, NewYork, 1981)
10. J.B. Pollack, O. Hubickyj, P. Bodenheimer, J.J. Lissauer, M. Podolak, Y. Greenzweig, Formation of the giant planets by concurrent accretion of solids and gas. Icarus **124**, 62–85 (1996)
11. A.P. Boss, Convective cooling of protoplanetary disks and rapid giant planet formation. ApJ **610**, 456–463 (2004)
12. R.M. Canup, W.R. Ward, Formation of the galilean satellites: conditions of accretion. AJ **124**, 3404–3423 (2002)
13. D. Jewitt, N. Haghighipour, Irregular satellites of the planets: products of capture in the early solar system. Ann. Rev. A&A **45**, 261–295 (2007)
14. C.B. Agnor, D.P. Hamilton, Neptune's capture of its moon Triton in a binary-planet gravitational encounter. Nature **441**, 192–194 (2006)
15. R.M. Canup, W.R. Ward, A common mass scaling for satellite systems of gaseous planets. Nature **441**, 834–839 (2006)
16. P. Podsiadlowski, S. Rappaport, J.M. Fregeau, R.A. Mardling, On the possibility of tidal formation of binary planets around ordinary stars (ArXiv e-prints, 2010)
17. J.W. Barnes, D.P. O'Brien, Stability of satellites around close-in extrasolar giant planets. ApJ **575**, 1087–1093 (2002)
18. F. Namouni, The fate of Moons of close-in giant exoplanets. ApJL **719**, L145–L147 (2010)
19. I.O. Yarkovsky, *Hypothese cinetique de la Gravitation universelle et connexion avec la formation des elements chimiques* (*Kinetic hypothesis of the universal gravitation and its connection with the formation of chemical elements*), 1st edn. (Chez lAuteur, Moscou, 1888)
20. R.C. Domingos, O.C. Winter, T. Yokoyama, Stable satellites around extrasolar giant planets. MNRAS **373**, 1227–1234 (2006)
21. J.R. Donnison, The Hill stability of inclined small mass binary systems in three-body systems with special application to triple star systems, extrasolar planetary systems and Binary Kuiper Belt systems. Planet. Space Sci. **58**, 1169–1179 (2010). doi:10.1016/j.pss.2010.04.009
22. G.J. Consolmagno, Ice-rich moons and the physical properties of ice. J. Phys. Chem. **87**, 4204–4208 (1983)

23. D. Valencia, R.J. O'Connell, D. Sasselov, Internal structure of massive terrestrial planets. Icarus **181**, 545–554 (2006)
24. C. Liebig, J. Wambsganss, Detectability of extrasolar moons as gravitational microlenses. A&A **520**, A68+ (2010)
25. K.M. Lewis, P.D. Sackett, R.A. Mardling, Possibility of detecting moons of pulsar planets through time-of-arrival analysis. ApJL **685**, L153–L156 (2008)
26. van W. Straten, M. Bailes, M. Britton, S.R. Kulkarni, S.B. Anderson, R.N. Manchester, J. Sarkissian, A test of general relativity from the three-dimensional orbital geometry of a binary pulsar. Nature **412**, 158–160 (2001)
27. A. Wolszczan, D.A. Frail, A planetary system around the millisecond pulsar PSR1257+12. Nature **355**, 145–147 (1992)
28. D.C. Backer, R.S. Foster, S. Sallmen, A second companion of the millisecond pulsar 1620–26. Nature **365**, 817–819 (1993)
29. R. Goullioud, J.H. Catanzarite, F.G. Dekens, M. Shao, J.C. Marr, IV. Overview of the sim planetquest light mission concept. society of photo-optical instrumentation engineers (SPIE) conference series, vol. 7013 of society of photo-optical instrumentation engineers (SPIE) conference series (2008). doi: 10.1117/12.789988
30. M.H.M. Morais, A.C.M. Correia, Stellar wobble caused by a binary system: can it really be mistaken as an extra-solar planet? A&A **491**, 899–906 (2008). doi: 10.1051/0004-6361:200810741
31. R. Martínez-Arnáiz, J. Maldonado, D. Montes, C. Eiroa, B. Montesinos, Chromospheric activity and rotation of FGK stars in the solar vicinity. An estimation of the radial velocity jitter. A&A **520**, A79+ (2010)
32. J.P. Beaulieu, D.P. Bennett, V. Batista, A. Cassan , D. Kubas, P. Fouqué, E. Kerrins, S. Mao, J. Miralda-Escudé, J. Wambsganss, B.S. Gaudi, A. Gould, S. Dong, *EUCLID: Dark Universe Probe and Microlensing Planet Hunter*. in ed. by V. Coudé Du Foresto, D.M. Gelino, I. Ribas. Astronomical Society of the Pacific Conference Series, Astronomical Society of the Pacific Conference Series, vol. 430 (2010), p. 266–+

Chapter 3
The Transiting Planet

From immemorial antiquity, men have dreamed of a royal road to success-leading directly and easily to some goal that could be reached otherwise only be long approaches and with weary toil. Times beyond number, this dream has proved to be a delusion... Nevertheless, there are ways of approach to unknown territory which lead surprisingly far, and repay their followers richly. There is probably no better example of this than eclipses of heavenly bodies

Henry Norris Russell, 1946

3.1 The Planetary Orbit

3.1.1 Orbital Elements

The first step in understanding the transit light curve is to define the appropriate coordinate system. In defining the so-called orbital elements, a Cartesian coordinate system is usually adopted. I begin with the simplest reference frame possible, where the planet orbits the star in the $\hat{x}$–$\hat{y}$ plane with the star at one focus, defined to be the origin, as shown in Fig. 3.1. By working in the rest frame of the star, the reflex motion is inherently accounted for by the model. Concordantly, the Cartesian coordinate of the planet may be expressed as a function of the planet's true anomaly, f_P:

$$\begin{aligned} x_P &= r_P \cos f_P \\ y_P &= r_P \sin f_P \\ z_P &= 0 \end{aligned} \tag{3.1}$$

Where r_P is the planet–star separation, given by:

$$r_P = a_P \frac{1 - e_P^2}{1 + e_P \cos f_P} = a_p \varrho_P(f_P) \tag{3.2}$$

D. M. Kipping, *The Transits of Extrasolar Planets with Moons*, Springer Theses,
DOI: 10.1007/978-3-642-22269-6_3, © Springer-Verlag Berlin Heidelberg 2011

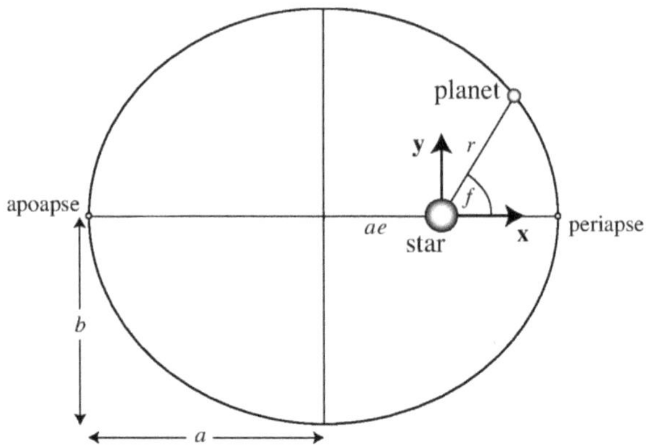

Fig. 3.1 The orbit of a planet around a star in the simplest reference frame. The orbit follows an elliptical path with the star at one focus, in accordance with Kepler's First Law. Figure adapted from Murray and Correia [2].

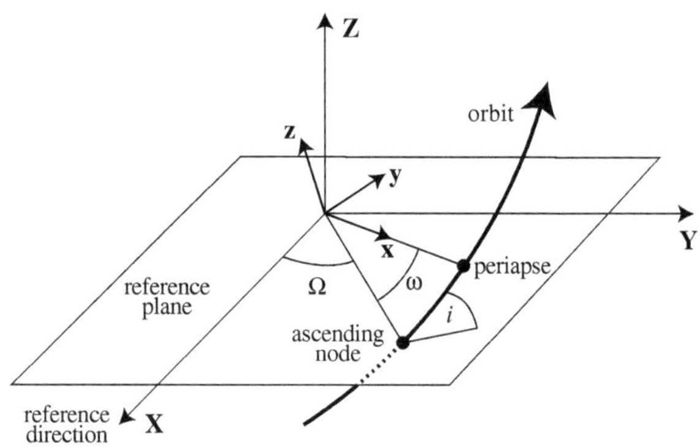

Fig. 3.2 Orbital elements of a planet orbiting a star. In the standard theory of transits, $i \simeq 90°$ and the observer is located at $\{X, Y, Z\} = \{0, 0, +\infty\}$. Figure from Murray and Correia [2].

Where $\varrho_P(f_P)$ is used to absorb the effects of orbital eccentricity. This simple 2D picture is, of course, impractical since real orbits exist in 3D. Say one executes N rotations in three dimensions to account for the viewing angle of the target solar system in its local cluster, then the rotation of that cluster to the Galactic plane and then the Solar System's rotation relative to the Galactic plane, etc. One can very quickly obtain a large integer value for N which would severely complicate the mathematics. However, Euler [1] showed that any finite chain of rotations can be simplified to just three rotations.

Standard celestial mechanics exploits Euler's theorem so that any planetary system can be described by just three angles: the argument of periapse, ω_P, the orbital inclination angle, i_P, and the longitude of the ascending node, Ω_P. The rotations are performed sequentially in a clockwise sense, about the axes $\hat{z}$–$\hat{x}$–$\hat{z}$. The rotations are shown in Fig. 3.2, which illustrates the orbital elements.

As one is performing a series of three dimensional rotations, the easiest way to represent the transformations mathematically is using matrices. Following [2], I denote the following clockwise rotation matrices about the $\hat{x}$ and $\hat{z}$ axes respectively:

$$\mathbf{P}_x(\phi) = \begin{pmatrix} 1 & 0 & 0 \\ 0 & \cos\phi & -\sin\phi \\ 0 & \sin\phi & \cos\phi \end{pmatrix} \tag{3.3}$$

$$\mathbf{P}_z(\phi) = \begin{pmatrix} \cos\phi & -\sin\phi & 0 \\ \sin\phi & \cos\phi & 0 \\ 0 & 0 & 1 \end{pmatrix}. \tag{3.4}$$

Armed with these definitions, the transformation from the simple frame of $\{x, y, z\}$ to the observed frame of $\{X, Y, Z\}$ may be written as:

$$\begin{pmatrix} X_P \\ Y_P \\ Z_P \end{pmatrix} = \mathbf{P}_z(\Omega_P)\mathbf{P}_x(i_P)\mathbf{P}_z(\omega_P) \begin{pmatrix} x_P \\ y_P \\ z_P \end{pmatrix} \tag{3.5}$$

After simplification, this yields the following final Cartesian coordinates:

$$\begin{aligned} X_P &= r_P[\cos\Omega_P \cos(\omega_P + f_P) - \sin\Omega_P \sin(\omega_P + f_P)\cos i_P] \\ Y_P &= r_P[\sin\Omega_P \cos(\omega_P + f_P) + \cos\Omega_P \sin(\omega_P + f_P)\cos i_P] \\ Z_P &= r_P \sin(\omega_P + f_P)\sin i_P \end{aligned} \tag{3.6}$$

In the final frame, plane of the sky is $\hat{X}$–$\hat{Y}$ and the observer is located far down the $+\hat{Z}$ axis at $\{0, 0, +\infty\}$. From these simple equations, one is able to derive a large proportion of the equations governing exoplanetary transit light curves. However, a description of the planetary motion is still missing.

3.1.2 Kepler's Equation

Equation (3.6) provides the instantaneous position of a planet as seen by the observer in three dimensions. However, one complication of the equation is that it is expressed in terms of the true anomaly of the planet, f_P, rather than time, t. To convert from f to t is non-trivial and requires solving Kepler's equation. A detailed discussion of the origins of the Kepler's equation can be found in [3], but for the purposes of this thesis, a statement of the relevant equations is sufficient. In standard notation, a new parameter known as the mean anomaly, $\mathfrak{M}$, is specified, which by definition scales linearly with time:

$$\mathfrak{M}(t) = nt + \mathfrak{M}_0 \tag{3.7}$$

The mean anomaly does not directly relate to the true anomaly but can be connected using an intermediary term known as the eccentric anomaly, E. The relationship between $\mathfrak{M}$ and E is given by:

$$\mathfrak{M} = E - e\sin E \tag{3.8}$$

Solving E as a function of $\mathfrak{M}$ gives a transcendental equation, which is a source of perennial nuisance in computing orbits. This is typically done with a numerical iteration, starting from $E_1 = \mathfrak{M}$, and I opted for the Newton–Rhapson iteration method [4], given by:

$$E_{i+1} = E_i - \frac{E_i - e \sin E_i - \mathfrak{M}}{1 - e \cos E_i} \tag{3.9}$$

Once E has been found to the desired level of precision, the true anomaly is computed using:

$$\tan\frac{f}{2} = \sqrt{\frac{1+e}{1-e}}\tan\frac{E}{2} \tag{3.10}$$

A very useful equation to introduce at this point is the rate of change of f, with respect to time:

$$\frac{\mathrm{d}t}{\mathrm{d}f} = \frac{P}{2\pi}\frac{(1-e^2)^{3/2}}{(1+e\cos f)^2} \tag{3.11}$$

3.1.3 Kepler's Laws

Kepler's Laws of planetary motion are one of the bedrocks underlying the physics of exoplanetary transits and it is appropriate to mention it here as part of the fundamentals of transit theory. The three laws are:

1. The orbit of every planet is an ellipse with the Sun at one of the two foci (i.e. Fig. 3.1)
2. A line joining a planet and the Sun sweeps out equal areas during equal intervals of time
3. The square of the orbital period of a planet is directly proportional to the cube of the semi-major axis of its orbit

Kepler's Third Law is particularly cogent for exoplanet transits, as will be seen later in Sect. 4.2.2. It was originally stated as $P^2 \sim a^3$ but Newton's version of Kepler's Third Law (which I simply refer to Kepler's Third Law in the rest of this thesis) provides the constants of proportionality, derived from his Laws of Gravitation. For two bodies labelled with subscripts 1 and 2, in the reference frame of body 1's rest frame, Kepler's Third Law states:

$$\left(\frac{P_1}{2\pi}\right)^2 = \frac{a_1^3}{G(M_1 + M_2)} \tag{3.12}$$

3.2 The Radial Velocity Signal

A useful demonstration of the power of the equations obtained thus far is illustrated in the derivation of the radial velocity signal of a transiting planet. As discussed in Sect. 1.3.5, radial velocity measures the reflex motion of a star due to the presence of a companion along the line-of-sight. One can now see that the line-of-sight is $\hat{Z}$. Let us denote the motion of the separation between the star and the planet–star barycentre as r_*. Since r_P is the planet–star separation then, $r_p = r_P M_*/(M_* + M_P)$ will be the planet-barycentre separation and finally $r_* = r_P M_P/(M_* + M_P)$ will be the star-barycentre separation. Therefore, we have:

$$r_* = a_P \varrho_P(f_P)\left(\frac{M_P}{M_* + M_P}\right) \tag{3.13}$$

In Sect. 3.1.1, i selected a reference frame with the star at rest, but here one is interested in the star's reflex motion. To move from the star's rest frame to that of the planet–moon barycentre is quite straightforward. I simply replace $r_P \rightarrow r_*$ in Eq. (3.6) and then make use of Eq. (3.13) to obtain the correct description of the star's observed reflex motion in three dimensions. For radial velocity, the relevant term is Z_*:

$$Z_* = a_P \varrho_P(f_P)\left(\frac{M_P}{M_* + M_P}\right) \sin(\omega_P + f_P) \sin i_P \tag{3.14}$$

Note that Z_* does not contain any dependency upon Ω_P and thus RV measurements cannot constrain this parameter. The RV method measures $\mathrm{d}Z_*/\mathrm{d}t$, but a minus sign is usually introduced to give the historically standard definition of a positive RV corresponding to a receding object:

$$\begin{aligned} \mathrm{RV} &= \gamma - \frac{\mathrm{d}Z_*}{\mathrm{d}t} \\ &= \gamma - \frac{\mathrm{d}Z_*}{\mathrm{d}f_P}\frac{\mathrm{d}f_P}{\mathrm{d}t} \end{aligned} \tag{3.15}$$

Where γ represents the constant systematic drift of the star relative to the Sun, along our line of sight. Differentiating Z_* with respect to f_P and substituting Eq. (3.11) into (3.15):

$$\mathrm{RV} = \gamma - K \sin i_P[e_P \cos\omega_P + \cos(f_P + \omega_P)] \tag{3.16}$$

Where I have substituted K, known as the "radial velocity semi-amplitude", for:

$$\begin{aligned} K &= a_P n_P \left(\frac{M_P}{M_* + M_P}\right)(1 - e_P^2)^{-1/2} \\ &= G^{1/3} n_P^{1/3} \left(\frac{M_P}{(M_* + M_P)^{2/3}}\right)(1 - e_P^2)^{-1/2} \end{aligned} \tag{3.17}$$

Where I have used Kepler's Third Law Eq. (3.12) on the second line. The mass terms may be grouped into a parameter sometimes referred to as the "mass function", given by $M_P(M_*+M_P)^{-2/3}$. If one is able to determine K from a set of radial velocity measurements, and the eccentricity, inclination, period and stellar mass are known, it should be possible to determine M_P. Solving Eq. (3.17) for M_P in this way yields a cubic with only one real root:

$$\begin{aligned} M_P = \Big[& -12GK^3M_*P_P\pi\xi_P^{3/2} - K^6P_P^2\xi_P^3 + K^3P_P\xi_P^{3/2}\Big(-54G^2K^3M_*^2P_P\pi^2\xi_P^{3/2} \\ & - 18GK^6M_*P_P^2\pi\xi_P^3 - K^9P_P^3\xi_P^{9/2} + 6\sqrt{3}\pi^{3/2} \\ & \times\sqrt{G^3K^6M_*^3P_P^2\xi_P^3(27GM_*\pi + 2K^3P_P\xi_P^{3/2})}\Big)^{1/3} \\ & - \Big(-54G^2K^3M_*^2P_P\pi^2\xi_P^{3/2} - 18GK^6M_*P_P^2\pi\xi_P^3 - K^9P_P^3\xi_P^{9/2} \\ & + 6\sqrt{3}\pi^{3/2}\sqrt{G^3K^6M_*^3P_P^2\xi_P^3(27GM_*\pi + 2K^3P_P\xi_P^{3/2})}\Big)^{2/3}\Big] \\ & \times\Big[6G\pi\Big(-54G^2K^3M_*^2P_P\pi^2\xi_P^{3/2} - 18GK^6M_*P_P^2\pi\xi_P^3 - K^9P_P^3\xi_P^{9/2} \\ & + 6\sqrt{3}\pi^{3/2}\sqrt{G^3K^6M_*^3P_P^2\xi_P^3(27GM_*\pi + 2K^3P_P\xi_P^{3/2})}\Big)^{1/3}\Big]^{-1} \end{aligned} \tag{3.18}$$

Where i have used $\xi_P = (1-e_P^2)$. I have not found this equation previously in the exoplanet literature but it is clearly preferable to use this exact solution as opposed to approximating $M_* \gg M_P$ in the mass function given in Eq. (3.17) and solving from there.

3.3 The Transit Light Curve for a Uniform Source Star

3.3.1 Transit Basics

We have now accumulated sufficient background knowledge to describe the transit light curve. For the purpose of introducing the topic, I will here only describe the case of a star of uniform brightness, which eliminates the complications of limb darkening effects. Although a detailed account of limb darkening is not given in this thesis for the sake of brevity, I will later briefly discuss in Sect. 3.3.6 how it is incorporated into my modelling routines.

I start by assuming that A1) there are only two bodies in the system A2) the motion of these bodies is completely described by Newton's laws only. The second assumption clearly excludes many other possible complications e.g. relativistic effects, Newton's laws are not universal, tidal dissipation, disc migration, etc. Armed with these two assumptions, the equations of motion provided earlier in Sect. 3.1 are valid here. I also define assumption A0 to be that the star has uniform brightness.

As was seen in Sect. 3.2, the star exhibits reflex motion, which means that in the inertial frame of the planet–star barycentre there are two bodies in motion. It is therefore simpler to choose a frame with the star at rest, as was done in Sect. 3.1.1, and so the position of the planet is given by Eq. (3.6). Since the star is at rest then, this equation also describes the planet-star separation, which is of course the relevant issue for transits.

More specifically, it is not really the planet–star separation which matters, but the *sky-projected* planet–star separation. In other words, I acknowledge that an observer has a certain perspective of the system. The sky-projected planet–star separation, S'_{P*}, is given by the quadrature sum of the two coordinates lying in the sky-plane[1]:

$$\begin{aligned} S'_{P*} &= \sqrt{X_P^2 + Y_P^2} \\ &= \frac{a_P(1 - e_P^2)}{1 + e_P \cos f_P} \sqrt{1 - \sin^2(\omega_P + f_P) \sin^2 i_P} \end{aligned} \tag{3.19}$$

S'_{P*} completely describes the transit light curve for a uniformly emitting star. Note how there is no dependence on Ω_P, just as we had for the RV signal, and so once again this parameter cannot be determined from the transit light curve. For a planet crossing over the stellar disc, there are three, and only three, distinct cases:

- *Out-of-transit*: No part of the sky-projected planetary disc overlaps with the sky-projected stellar disc
- *On-the-limb*: The sky-projected planetary disc lies partially within the sky-projected stellar disc
- *In-transit*: The sky-projected planetary disc lies fully inside the sky-projected stellar disc

So far I have only made three assumptions; A0, A1 and A2. Let us make the additional assumption, A3, that both the planet and star are perfect spheres. The 2D projection of a sphere viewed from any angle will always appear as a circle, which means that the three cases mentioned above can be completely described in terms of the radii of these circles (radius of the planet, R_P, and radius of the star, R_*) and S'_{P*}.

- *Out-of-transit*: $R_* + R_P \leq S'_{P*} < \infty$
- *On-the-limb*: $R_* - R_P \leq S'_{P*} < R_* + R_P$
- *In-transit*: $0 \leq S'_{P*} < R_* - R_P$

With the three cases defined, let us ask what is the observed flux for each case. To do this, one has to make several new assumptions which are often not explicitly stated. However, for completion, I will here state all of these assumptions (including those stated earlier):

[1] The subscripts "*P*" and "*" in S'_{P*} are commutative

A0 The star has uniform brightness
A1 There are only two bodies in the system
A2 The motion of these bodies is completely described by Newton's laws only
A3 Both the planet and star are perfect spheres
A4 The planet emits no flux and is completely opaque
A5 The star's emission is of constant flux
A6 There are no background/foreground luminous objects
A7 The exoplanetary system is a constant distance d from the observer, such that $d \gg a_P$
A8 The planet has no extended features, such as an atmosphere, rings, etc
A9 The planet does not cause gravitational microlensing of the host star's light
A10 The change in flux over a single integration is much smaller than the flux measurement uncertainty

Now, that in-transit, the flux of a star is attenuated from $F_* \rightarrow F_*[1 - (R_P^2/R_*^2)]$ i.e. the ratio-of-areas. Therefore, the depth of a transit, δ, is:

$$\delta = R_P^2/R_*^2 = p^2 \tag{3.20}$$

Where I have used $p = R_P/R_*$ i.e. the ratio-of-radii. As p is essentially a measurement of the planetary radius in units of the stellar radius, it will be useful to translate S'_{P*} to also be in units of the stellar radius. I therefore define $S_{P*} = S'_{P*}/R_*$. I now normalize the observed flux at any instant to that occurring during the out-of-transit times, to give the so-called "normalized flux", F_*^N.

$$F_{*,\text{transit}}^N(S_{P*}) = \begin{cases} 1 & 1 + p \leq S_{P*} < \infty \\ 1 - \alpha(1, p; S_{P*}) & 1 - p \leq S_{P*} < 1 + p \\ 1 - p^2 & 0 \leq S_{P*} < 1 - p \end{cases} \tag{3.21}$$

Where $\alpha(R, r; S)$ is the area of overlap between two circles, of radii R and r with separation S. This simple scenario has a well-known solution:

$$\alpha(R, r; S) = r^2\kappa_0(R, r; S) + R^2\kappa_1(R, r; S) - \kappa_2(R, r; S) \tag{3.22}$$

$$\kappa_0(R, r; S) = \arccos\left[\frac{S^2 + r^2 - R^2}{2Sr}\right] \tag{3.23}$$

$$\kappa_1(R, r; S) = \arccos\left[\frac{S^2 + R^2 - r^2}{2SR}\right] \tag{3.24}$$

$$\kappa_2(R, r; S) = \sqrt{\frac{4S^2R^2 - (R^2 + S^2 - r^2)^2}{4}} \tag{3.25}$$

3.3.2 Conditions for a Transit

One can see from Eq. (3.21) that if there is no instance in the orbit where $S_{P*} < 1+p$, then $F_*^N = 1$ at all times and so no transit features will be observable. Therefore, the condition for a transit to occur is dependent upon the minimum value of S_{P*}. Let us denote this minimum as $S_{P*,T}$ and the true anomaly at which it occurs as $f_{P,T}$ (the "T" subscript denotes transit, as opposed to "O" for occultation which is discussed more in Sect. 3.3.4).

$S_{P*,T}$ occurs approximately at the time of inferior conjunction, but the exact solution has a rather elaborate analytic form and is presented in Sect. 4.1. Proceeding with the approximate form for now, the minimum corresponding to the primary transit is therefore $f_{P,T} \simeq (\pi/2 - \omega_P)$. Substituting $f_{P,T}$ into Eq. (3.19), one obtains $S_{P*,T}$:

$$S_{P*,T} \simeq \frac{a_P}{R_*} \frac{1 - e_P^2}{1 + e_P \sin \omega_P} \cos i_P = b_{P,T} \tag{3.26}$$

Where $b_{P,T}$ is known as the "impact parameter" of the planetary transit event.[2] Therefore, for a full-transit to occur one requires $b_{P,T} \leq 1 - p$ and for a so-called "grazing" transit one requires $(1-p) < b_{P,T} < (1+p)$. For a circular orbit, one can see that two critical parameters will impinge on the light curve, $b_{P,T}$ and a_P/R_*. For an eccentric orbit, one has the additional parameter $\varrho_{P,T} = \varrho_P(f_P = \pi/2 - \omega_P)$ (see Eq. (3.2) for the definition of $\varrho_P(f_P)$). The importance of these parameters becomes apparent when fitting light curves.

3.3.3 Anatomy of the Transit Light Curve

As established in the previous section and stated explicitly in Eq. (3.21), there are three distinct cases for the transit. The three cases are buffeted by two boundaries i.e. $S_{P*} = (1 + p)$ and $S_{P*} = (1 - p)$. For a grazing transit, only the former of these two boundaries is crossed, but for a full-transit both are traversed. In either case, the planet must both enter and exit these boundaries and thus a full-transit has a total of four boundary crossings, which are usually labelled as the "contact points" of the transit.

These contact points are labelled sequentially as times t_I, t_{II}, t_{III} and t_{IV}, and the instant when S_{P*} is minimized is τ_T (see Fig. 3.3). The durations between these contact points will be discussed in Chap. 4.

[2] Some of the subscripts present in the notation employed at this stage of the thesis may seem superfluous, but once moons and occultations are introduced, the value of these subscripts will become apparent

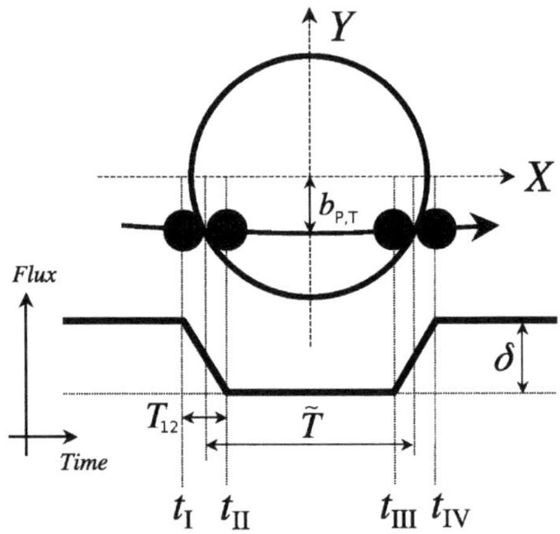

Fig. 3.3 Anatomy of the transit light curve. The four key contact points are labelled along the bottom and the corresponding position of the planet is visible from the upper part of the figure. The transit durations $\tilde{T}$ and T_{12} are also marked, which will be discussed further in Chap. 4 The impact parameter, $b_{P,T}$, is defined as the sky-projected planet-star separation in units of the stellar radius at the instant of inferior conjunction. Figure adapted from Winn [47].

3.3.4 Occultations

As well as the primary transit event, secondary eclipses or "occultations" can occur too. Occultations cause a much shallower eclipse depth as one is now seeing the flux decrease from $(F_* + F_P) \rightarrow F_*$ and, in general, $F_P \ll F_*$ (where F_P is the flux from the planet). F_P can be caused by reflected light and thermal emission from the exoplanet and so in both cases it is largest for close-in planets. By definition, in order to acknowledge the presence of an occultation one must break assumption A4. This does affect the transit event and will be discussed in Sect. 5.2. Proceeding with the occultation for now, one may combine the two possible sources for the eclipse, to calculate the occultation depth, ϵ:

$$\epsilon = \left[\underbrace{\frac{\int_{\lambda=0}^{\infty} R_I(\lambda)\mathfrak{A}_G(\lambda)\mathrm{d}\lambda}{(a_P/R_*)^2}}_{\text{reflection}} + \underbrace{\frac{\int_{\lambda=0}^{\infty} R_I(\lambda)F_P(\lambda)\mathrm{d}\lambda}{\int_{\lambda=0}^{\infty} R_I(\lambda)F_*(\lambda)\mathrm{d}\lambda}}_{\text{thermal emission}} \right] p^2 \qquad (3.27)$$
$$= \mathfrak{F}_{P*}p^2$$

Where the left-hand side is for reflected light and the right-hand side for thermal emission. $R_I(\lambda)$ denotes the response function of the instrument used to observe the star. All of these effects can be absorbed into the quantity $\mathfrak{F}_{P*}$, the ratio of the planet's flux per unit area to that of the star. This means in modelling the occultation one may use:

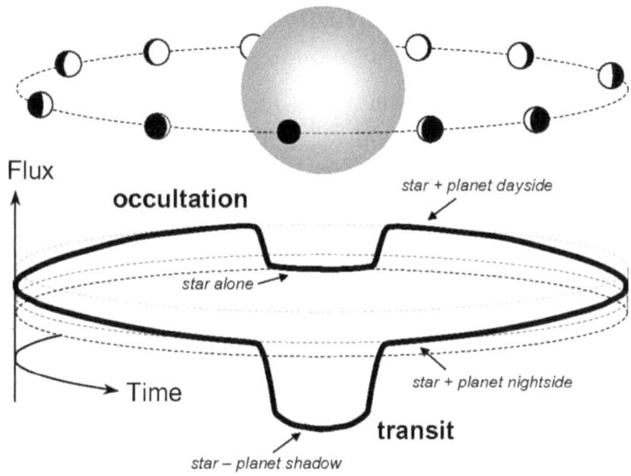

Fig. 3.4 Illustration of the phase curve of an exoplanet in the context of the primary and secondary transits. As the day-side comes into view, the flux from the planet appears to increase giving rise to a slight increase in the total flux observed from the star+planet pair. Figure from Winn [47]

$$F^N_{*,\text{eclipse}}(S_{P*}) = \begin{cases} 1 & 1+p \leq S_{P*} < \infty \\ 1 - \mathfrak{F}_{P*}\alpha(1, p; S_{P*}) & 1-p \leq S_{P*} < 1+p \\ 1 - \mathfrak{F}_{P*}p^2 & 0 \leq S_{P*} < 1+p \end{cases} \tag{3.28}$$

Just as transits occur near the time of inferior conjunction, occultations occur near the time of superior conjunction. For circular orbits, the presence of a primary transit guarantees that an occultation of equal duration must occur at a time $\tau_O = \tau_T + 0.5P_P + \Delta t_{\text{light}}$, where Δt_{light} is the light travel time across the system and is typically less than a minute for hot-Jupiters [5]. However, for eccentric orbits the presence of an occultation is not guaranteed and the duration and time of the event may also vary, which can be exploited to actually measure the eccentricity [6]. These points will be addressed in Sect. 4.1 and Sect. 4.4.1.

3.3.5 Phase Curves

For a hot-Jupiter system, the very short-period is thought to cause tidal locking of the planets rotational period to the orbital period [7]. Therefore, the planet has the same side facing the star at all times leading to potentially large changes in emitted flux between the day and night side. As the planet moves from the primary to secondary positions, one sees the day side come into view and thus continuous monitoring could reveal small changes in the total flux, known as the "phase curve", indicative of the day–night contrast (the first successful observation of this effect was achieved by Knutson et al. [8]). Figure 3.4 provides an illustration of this in the context of the primary and occultations.

3.3.6 Extending to Limb Darkened Stars

The equations presented thus far allow one to fully model the primary transit and occultations of an exoplanet. Whilst occultations are completely unaffected by stellar limb darkening, the primary light curve is modified. As limb darkening is circularly symmetric, the S_{P*} parameter is still the only term one needs to describe the light curve. This is contrast to gravity darkening, which breaks this symmetry and thus requires one to compute X_P and Y_P separately at every instance. Fortunately, gravity darkening is generally ineffective and may be neglected, although it does become significant for rapidly rotating stars (spectral types ~F6 and earlier) and the appropriate treatment is described in Barnes [9].

Limb darkened stars exhibit a monotonically decreasing brightness profile from centre to limb. Numerous profile forms have been proposed including linear [10], quadratic [10], square-root [10], four-coefficient polynomial [11] and logarithmic [11]. Claret [11] argue that the four-coefficient model gives the most accurate representation but the quadratic model is sufficient for most applications. The use of just two coefficients rather than four is preferable from a parameter fitting perspective and I will adopt the quadratic law in this thesis. I treat the specific stellar intensity as:

$$\frac{I_\mu}{I_1} = 1 - u_1(1 - \mu) - u_2(1 - \mu)^2 \tag{3.29}$$

Where $\mu \equiv \cos\theta$, θ is the angle between the stellar surface normal vector and the line-of-sight, and u_1 and u_2 are the linear and quadratic limb darkening coefficients respectively.

Mandel and Agol [12] presented equations for modelling the transit light curve with linear, quadratic and four-coefficient polynomial limb darkening. Another benefit of the quadratic model is that Mandel and Agol [12] find the four-coefficient method requires computing hypergeometric functions, which slow down a fitting routine significantly. In this thesis, I employ the quadratic expressions from [12] to generalize the modelling of transit light curves and thus assumption A0 (a uniform source) is broken.

3.3.7 Breaking the Assumptions

Breaking any one of the assumptions listed in Sect. 3.1.1 will invalidate the equations presented. As one may expect, some of the assumptions hold in almost all situations whereas others are almost invariably broken. The consequences and likelihood of each assumption being broken is discussed here

3.3.7.1 A0) A uniformly emitting star

I have discussed how limb darkening [12] strongly affects the transit light curve but has no effect on the occultation event. Limb darkening becomes less prominent for stars observed in the infrared wavelengths but in general is always present and so it is prudent to always account for limb darkening.

Gravity darkening [9] is rarely prominent and currently no light curves of a transiting planet have ever been shown to exhibit this effect. However, rapidly rotating stars will have significant gravity darkening and in these cases one should model the effect [9].

Other non-uniform effects, such as starspots and plages, can have a significant effect on the transit light curve, especially at visible wavelengths. If a planet transits across a starspot then an apparent flux increment will occur which may be used to determine the spot's parameters [13]. Much more troubling (and more probable) are out-of-transit spots which can induce variations in the transit depth from epoch-to-epoch [14].

3.3.7.2 A1) There are only two bodies in the system

For just two bodies following Keplerian orbits, the orbital motions are described with Eq. (3.11). Accordingly, transits occur once every orbital period, P_P. However, the introduction of a third body leads to an extremely challenging analytic problem. Indeed, the general solution to the three-body problem has never been found and as such one is forced to use either approximate solution for restricted cases or numerical methods.

Additional bodies, whether a nearby star, outer/inner planet or companion moon will perturb the orbital motion of the transiting planet. More specifically, the gravitational influence of the third body acts to perturb the position and velocity of the transiting body, which manifest as transit time and duration variations (TTV and TDV). TTV means that the planet no longer transits once every orbital period (the deviation can be seconds up to days) and TDV means that the duration of the transit events vary from orbit-to-orbit.

Agol et al. [15] and Holman and Murray [16] discuss the TTV effects of a perturbing planet and show that bodies in mean motion resonance (MMR) with the transiting planet induce heavily amplified signals. Borkovits et al. [17] recently extended these analyses to longer time scale perturbations too. Apsidal and nodal precession which may be induced by a third body lead to TTV and TDV effects as well [18, 19]. Sartoretti and Schneider [20] discuss the TTV effects caused by a companion moon and Kipping [19, 46] predict an accompanying TDV signal (also see Chap. 6). Ford and Holman [21] discuss the TTV effect caused by a Trojan body.

Aside from TTV and TDV, the third body can also transit the host star as well. If these transits are simultaneous (e.g. the case of a moon), then the transit signal can modify the light curve shape in an elaborate manner. Whilst authors such as Sato and Asada [22] have touched on these points, a generalized model which inherently includes both transit timing effects and the light curve features remains absent in the exoplanet literature. A new framework to solve this problem, `LUNA`, is currently under development (see Chap. 8).

3.3.7.3 A2) The motion are completely described by newton's laws only

Newton's laws produce very accurate predictions of the planetary motions in the Solar System. However, we have already seen in Chap. 1 how Mercury's observed precession rate cannot be explained by Newtonian mechanics alone [23]. General relativistic effects perturb the precession rate and so in this case one of the assumptions is clearly invalid. Jordán and Bakos [18] provide a detailed discussion of the observability of general relativistic precession and conclude that decades of duration measurements may be able to infer the rate. However, for time scales of years or less these effects are not significant.

Other effects include tidal dissipation and disc migration. Migration generally occurs on the time scale of Myr and thus is insignificant. Tidal dissipation is usually a slow process as well, although it has been hypothesized that in some extreme cases for planets on ultra-short periods it may result in TTV and TDV effects as the planet spirals inwards [24, 19]. Finally, stellar quadrupole moments may also cause timing deviations through the Applegate effect, but are generally long-term deviations ([25]).

3.3.7.4 A3) Both the planet and star are perfect spheres

Whilst a detailed discussion of stellar oblateness is absent in the exoplanet literature, planetary oblateness has been investigated. Hui and Seager [26] suggested such planets would induce asymmetric light curves but requiring photometric precisions at or beyond the ppm level. Carter and Winn [27] suggested that precession of the axial tilt of an oblate planet may offer a more detectable signal due to resultant depth changes, but predict decades of data are required in most cases. In general, these effects do not need to be included in a transit model.

3.3.7.5 A4) The planet emits no flux and is completely opaque

Although I could find no work considering the effects of a non-opaque planet, it is perhaps safe to assume that such an event is quite unlikely. For the planetary flux issue, occultations cannot be understood at all unless A4 is discarded. With many observations of such events, particularly at infrared wavelengths, now existing in the literature (e.g. Deming et al. [28], Charbonneau et al. [29], Swain et al. [30]) the assumption must be seen as frequently invalid.

Kipping and Tinetti [31] presented a method for correcting transit light curves for the effects of nightside emission (or "pollution" as dubbed by the authors due to the observational consequences) and this will be covered in this thesis in Sect. 5.2. As will be seen later, the overall effect is a predictable dilution of the transit depth.

3.3.7.6 A5) The star's emission is of constant flux

Stellar activity is always present at various degrees of intensity and depends heavily on the spectral type and rotation. Stars with very large activities are generally not selected for transit surveys and thus few examples exist in the literature. Those exhibiting pronounced activity induce photometric variations at the level of a percent which can induce TTV [32]. TDV is also predicted to be induced by magnetic activity in the star over the time scales of decades [33]. Techniques such as Fourier filtering [34] and wavelet analysis [35] can remove the bulk of the activity but of course it is preferable to select a "quiet" star in the first place (criterion C6).

Fortunately, 65–70% of F7-K9 main-sequence stars in the are likely to have similar or lower intrinsic variability than the Sun [36], which is 10 ppm. Such a tiny noise level is generally inconsequential to light curve analyses.

3.3.7.7 A6) There are no background luminous objects

Background luminous objects cause a dilution of transit depths, in a very similar way to how nightside emission does [31]). These dilutions are usually called "blends" and thus nightside emission is a self-blending scenario. More generally, blends mean that instead of seeing just F_* one has $F_* + F_B$, where F_B is the flux from the blended object. This causes the observed transit depth to attenuate to:

$$\delta' = \frac{\delta}{B} \tag{3.30}$$

$$B = \frac{F_* + F_B}{F_*} \tag{3.31}$$

Where B is the blending factor. Due to the simple nature of the effects of blends, it is easy to account for them in transit modelling if B is known. For surveys like *Kepler*, adaptive optics follow-up of transit candidates allows one to spatially resolve nearby stars from the target and thus determine B to high confidence (e.g. Latham et al. [37]). There is always a small chance that a blended target is too close to spatially resolve but the low probability of this occurrence is quantifiable and manageable.

The assumption of no luminous background objects also excludes gravitational microlensing of a distant background source. Such an effect would be extremely pronounced and easily identified but no examples of a transiting system acting as a lens star are known to exist. Lewis [38] estimate a very low probability, $\sim 10^{-6}$, of such an event being detected.

3.3.7.8 A7) The exoplanetary system is a large and constant distance Away

By treating the distance of the system, d, to be at a large distance away (i.e. $d \gg a_P$), one can neglect the angular size of the planet. The exact size of the shadow of the planet cast onto the stellar surface has a radius R'_P and causes a transit depth δ':

$$\begin{aligned} R'_P &= \frac{r_{P,T} + d}{d} R_P \\ \delta' &\simeq \left[1 + 2.34 \times 10^{-13} \left(\frac{a_P}{\mathrm{AU}} \right)^2 \left(\frac{10\mathrm{pc}}{d} \right)^2 \right] \delta \end{aligned} \tag{3.32}$$

Where $r_{P,T}$ is the planet–star separation at the transit minimum and on the second line I have approximated the orbit to be circular. The change in the transit depth is clearly negligible and thus assumption A7 seems justified.

Finally, one interesting effect which can occur for systems with high proper motion is a deviation in the times of transit due to the varying light travel time [39].

3.3.7.9 A8) The planet has no extended features

Planetary rings will induce transit features before and after the transit event. These features can be either brightenings due to forward scattering or dimmings due to opacity, but no detection of such phenomena have been made. Barnes and Fortney [40] provide simulations of transit light curves featuring ring systems and argue that *Kepler* should be able to detect many feasible configurations.

An atmosphere is also neglected in this analysis. The opacity of an atmosphere will vary as a function of wavelength due to absorption of light due by different molecules. This has been exploited to detect or constrain the presence of molecules in exo-atmospheres, in what is usually dubbed as "transmission spectroscopy" (e.g. Charbonneau et al. [41], Tinetti et al. [42], Beaulieu et al. [43]). For a single bandpass observation, this actually requires no modification to the model as R_P/R_* simply increases or decreases.

One atmospheric effect which can cause a breakdown of this model is refraction through the atmosphere, causing a lensing effect [26, 44]. The most significant perturbation is the presence of out-of-transit features caused by the lensing, which may mimic rings in their shape and size. No observations of these effects have been reported though, and they do not need to be included in general.

3.3.7.10 A9) The planet does not cause gravitational microlensing

Kasuya et al. [45] discuss the possibility of a planet acting as a gravitational lens to the host star and causing a degree of microlensing. Notably, spikes before and after the transit event may be observable, with amplitudes ~100 ppm. However, typically these events are only observable for transiting planets at very large separation from their host star ($\sim$200 AU), which means they are geometrically highly unlikely to transit in the first place ($\sim 10^{-5}$). Therefore, I do no consider this effect to be necessary to model in most circumstances.

3.3.7.11 A10) The change in flux over a single integration is much smaller than the flux measurement uncertainty

Taking long integrations for the photometric observations can result in a smearing of the transit signal. This effect was first discussed in [46] and I will come back to it in Sect. 5.3. The main lesson is that it is possible to modify the light curve model to incorporate the integration times and thus produce accurate parameter estimates for the system. One minute integration typically exhibit negligible smearing whilst anything above $\sim$15 min can cause significant effects. Given that the *Kepler Mission* performs the vast majority of its photometry using 30 min integrations, this effect can be quite severe and should be accounted for.

3.4 Summary

I have presented the fundamental equations for the sky-projected motion (Sect. 3.1), the associated radial velocity signal (Sect. 3.2) and observed light curve of a transiting planet (Sect. 3.3). The light curve for a uniform brightness star can be completely understood in terms of the sky-projected planet–star separation, S_{P*}, and this critical parameter is derived for two bodies in Keplerian orbits. The derived equations are easily manipulated to give the associated RV signal, which is critical in global modelling of the available data. Finally, I have presented a simple model for the observed flux during a transit event and clearly outlined my assumptions. The effect(s) of breaking the various assumptions are overviewed in Sect. 3.3.7 and I discuss the frequency and significance of each assumption being broken. In particular, I find that three out of the eleven assumptions are likely to be frequently broken with significant effects. These are A0 a uniform source star; A4/A5 the planet emits no flux/there are no background luminous objects; A10 the integration times are small. A4 and A5 are grouped together because they may be modelled using the same method, the introduction of a blending factor, B, which will be discussed in more detail in Sect. 5.2. For A0, I discuss how the expressions of Mandel and Agol [12] are used in my fitting routines to account for limb darkening. For A10, this assumption is broken by the long-cadence data of the *Kepler Mission* which uses 30 min integrations inducing smearing of the light curve. In Sect. 5.3, I will discuss a method to compensate for this effect.

With the basics established, there are numerous subtleties which I have skimmed over and wish to return to. Notably, the "transit time" is defined as the moment when S_{P*} is minimized and so far I have only stated that these moments occur near the times of inferior and superior conjunction for the transit and occultation events, respectively. Also, the duration of a transit event has not been addressed. All of the time-related quantities will be discussed in the next Chap. 4.

References

1. L. Euler, Novi Commentarii Academiae Scientiarum, 1st edn. Petropolitanae, 1776
2. C.D. Murray, A.C.M Correia, Keplerian Orbits and Dynamics. ArXiv e-prints, 2010
3. C.D. Murray, S.F Dermott, Solar system dynamics. (Cambridge University Press, Cambridge, 1999)
4. J.M.A. Danby, Fundamentals of celestial mechanics. (Willmann-Bell, Virginia, 1988)
5. A. Loeb, A dynamical method for measuring the masses of stars with transiting planets. Astrophys. J. Lett **623**, L45–L48 (2005)
6. D. Deming, J. Harrington, G. Laughlin, S. Seager, S.B. Navarro, W.C. Bowman, K. Horning, Spitzer transit and secondary eclipse photometry of GJ 436b. Astrophys. J. Lett. **667**, L199–L202 (2007)
7. B. Gladman, D.D. Quinn, P. Nicholson, R. Rand, Synchronous locking of tidally evolving satellites. Icarus **122**(1), 166–192 (1996)
8. H.A. Knutson, D. Charbonneau, L.E. Allen, J.J. Fortney, E. Agol, N.B. Cowan, A.P. Showman, C.S. Cooper, S.T. Megeath, A map of the day-night contrast of the extrasolar planet HD 189733b. Nature **447**, 183–186 (2007)
9. J.W. Barnes, Transit lightcurves of extrasolar planets orbiting rapidly rotating stars. Astrophys. J. **705**, 683–692 (2009)
10. A. Claret, J. Diaz-Cordoves, A. Gimenez, Linear and non-linear limb-darkening coefficients for the photometric bands R I J H K. Astron. Astrophys. Suppl. 114:247–+ (1995)
11. A. Claret, A new non-linear limb-darkening law for LTE stellar atmosphere models calculations for - 0 <= log[M/H] <= +1, 2000 K <= Teff <= 50000 K at several surface gravities.. Astron. Astrophys. Suppl. **363**, 1081–1190 (2000)
12. K. Mandel, E. Agol, Analytic light curves for planetary transit searches. Astrophys. J. Lett **580**, L171–L175 (2002)
13. F. Pont, R. L. Gilliland, C. Moutou, D. Charbonneau, F. Bouchy, T. M. Brown, M. Mayor, D. Queloz, N. Santos, S. Udry, Hubble space telescope time-series photometry of the planetary transit of HD 189733: no moon, no rings, starspots. Astron. Astrophys. **476**, 1347–1355 (2007)
14. S. Czesla, K.F. Huber, U. Wolter, S. Schröter, J.H.M.M Schmitt, How stellar activity affects the size estimates of extrasolar planets. Astron. Astrophys. **505**, 1277–1282 (2009)
15. E. Agol, J. Steffen, R. Sari, W. Clarkson, On detecting terrestrial planets with timing of giant planet transits. MNRAS **359**, 567–579 (2005)
16. M.J. Holman, N.W Murray, The use of transit timing to detect terrestrial-mass extrasolar planets. Science **307**, 1288–1291 (2005)
17. T. Borkovits, S. Csizmadia, E. Forgács-Dajka, T. Hegedüs, Transit timing variations in eccentric hierarchical triple exoplanetary systems. I. Long-term (P_2-time-scale) perturbations. ArXiv e-prints, October 2010.
18. A. Jordán, G.Á Bakos, Observability of the general relativistic precession of periastra in exoplanets. Astrophys. J. **685**, 543–552 (2008)
19. D.M. Kipping, Investigations of approximate expressions for the transit duration. MNRAS **407**, 301–313 (2010)
20. P. Sartoretti, J. Schneider, On the detection of satellites of extrasolar planets with the method of transits. Astron. Astrophys. Suppl. **134**, 553–560 (1999)
21. E.B. Ford, M.J Holman, Using transit timing observations to search for trojans of transiting extrasolar planets. Astrophys. J. Lett. **664**, L51–L54 (2007)

22. M. Sato, H. Asada, Effects of mutual transits by extrasolar planet-companion systems on light curves. PASJ. 61:L29+ (2009)
23. G.M. Clemence, The relativity effect in planetary motions. Rev. Mod. Phys. **19**, 361–364 (1947)
24. C. Hellier, D.R. Anderson, A.C. Cameron, M. Gillon, L. Hebb, P.F.L. Maxted, D. Queloz, B. Smalley, A.H.M.J. Triaud, R.G. West, D.M. Wilson, S.J. Bentley, B. Enoch, K. Horne, J. Irwin, T.A. Lister, M. Mayor, N. Parley, F. Pepe, D.L. Pollacco, D. Segransan, S. Udry, P.J Wheatley, An orbital period of 0.94 days for the hot-Jupiter planet WASP-18b. Nature **460**, 1098–1100 (2009)
25. C.A. Watson, T.R Marsh, Orbital period variations of hot Jupiters caused by the Applegate effect. MNRAS **405**, 2037–2043 (2010)
26. L. Hui, S. Seager, Atmospheric lensing and oblateness effects during an extrasolar planetary transit. Asrophys. J. **572**, 540–555 (2002)
27. J.A. Carter, J.N. Winn, The detectability of transit depth variations due to exoplanetary oblateness and spin precession. Asrophys. J. **716**, 850–856 (2010)
28. D. Deming, J. Harrington, S. Seager, L.J. Richardson, Strong infrared emission from the extrasolar planet HD 189733b. Asrophys. J. **644**, 560–564 (2006)
29. D. Charbonneau, H.A. Knutson, T. Barman, L.E. Allen, M. Mayor, S.T. Megeath, D. Queloz, S. Udry, The broadband infrared emission spectrum of the exoplanet HD 189733b. Asrophys. J. **686**, 1341–1348 (2008)
30. M.R. Swain, G. Tinetti, G. Vasisht, P. Deroo, C. Griffith, J. Bouwman, P. Chen, Y. Yung, A. Burrows, L.R. Brown, J. Matthews, J.F. Rowe, R. Kuschnig, D. Angerhausen, Water, methane, and carbon dioxide present in the dayside spectrum of the exoplanet HD 209458b. Asrophys. J. **704**, 1616–1621 (2009)
31. D.M. Kipping, G. Tinetti, Nightside pollution of exoplanet transit depths. MNRAS **407**, 2589–2598 (2010)
32. R. Alonso, M. Auvergne, A. Baglin, M. Ollivier, C. Moutou, D. Rouan, H.J. Deeg, S. Aigrain, J.M. Almenara, M. Barbieri, P. Barge, W. Benz, P. Bordé, F. Bouchy, R. deLaReza, M. Deleuil, R. Dvorak, A. Erikson, M. Fridlund, M. Gillon, P. Gondoin, T. Guillot, A. Hatzes, G. Hébrard, P. Kabath, L. Jorda, H. Lammer, A. Léger, A. Llebaria, B. Loeillet, P. Magain, M. Mayor, T. Mazeh, M. Pätzold, F. Pepe, F. Pont, D. Queloz, H. Rauer, A. Shporer, J. Schneider, B. Stecklum, S. Udry, G. Wuchterl, Transiting exoplanets from the CoRoT space mission. II. CoRoT-Exo-2b: a transiting planet around an active G star. Astron. Asrophys. **482**, L21–L24 (2008)
33. A. Loeb, Long-term evolution in transit duration of extrasolar planets from magnetic activity in their parent stars. New Astron. **14**, 363–364 (2009)
34. A. Alapini, S. Aigrain, An iterative filter to reconstruct planetary transit signals in the presence of stellar variability. MNRAS **397**, 1591–1598 (2009)
35. J.A. Carter, J.N. Winn, Parameter estimation from time-series data with correlated errors: a wavelet-based method and its application to transit light curves. Astrophys. J. **704**, 51–67 (2009)
36. N.M. Batalha, J. Jenkins, G.S. Basri, W.J. Borucki, D.G. Koch, Stellar variability and its implications for photometric planet detection with Kepler. In: B. Battrick, F. Favata, I. W. Roxburgh, D Galadi (eds.) Stellar Structure and Habitable Planet Finding. volume 485 of ESA, pp. 35-40, Special Publication, January 2002
37. D.W. Latham, W.J. Borucki, D.G. Koch, T.M. Brown, L.A. Buchhave, G. Basri, N.M. Batalha, D.A. Caldwell, W.D. Cochran, E.W. Dunham, G. Fűrész, T.N. Gautier, J.C. Geary, R.L. Gilliland, S.B. Howell, J.M. Jenkins, J.J. Lissauer, G.W. Marcy, D.G. Monet, J.F. Rowe, D.D Sasselov, Kepler-7b: a transiting planet with unusually low density. Astrophys. J. Lett. **713**, L140–L144 (2010)
38. G.F. Lewis, Gravitational microlensing of stars with transiting planets. Astron. Astrophys. **380**, 292–299 (2001)
39. R.R. Rafikov, Stellar proper motion and the timing of planetary transits. Asrtophys. J. **700**, 965–970 (2009)

40. J.W. Barnes, J.J. Fortney, Transit detectability of ring systems around extrasolar giant planets. Asrtophys. J. **616**, 1193–1203 (2004)
41. D. Charbonneau, T.M. Brown, R.W. Noyes, R.L. Gilliland, Detection of an extrasolar planet atmosphere. Astrophys. J. **568**, 377–384 (2002)
42. G. Tinetti, A. Vidal-Madjar, M. -C. Liang, J. -P. Beaulieu, Y. Yung, S. Carey, R.J. Barber, J. Tennyson, I. Ribas, N. Allard, G.E. Ballester, D.K. Sing, F. Selsis, Water vapour in the atmosphere of a transiting extrasolar planet. Nature. **448**, 169-171
43. J.P. Beaulieu, D.M. Kipping, V. Batista, G. Tinetti, I. Ribas, S. Carey, J.A. Noriega-Crespo, C.A. Griffith, G. Campanella, S. Dong, J. Tennyson, R.J. Barber, P. Deroo, S.J. Fossey, D. Liang, M.R. Swain, Y. Yung, N. Allard, Water in HD 209458b's atmosphere from 3.6 - 8 microns IRAC photometric observations in primary transit. ArXiv e-prints, September 2009
44. O. Sidis, R. Sari, Transits of Transparent Planets - Atmospheric Lensing Effects. ArXiv e-prints, August 2010
45. S. Kasuya, M. Honda, R. Mishima, New Observablefor Gravitational Lensing Effects During Transits. ArXiv e-prints, September 2010
46. D.M. Kipping, Binning is sinning: morphological light-curve distortions due to finite integration time. MNRAS **408**, 1758–1769 (2010)
47. J.N. Winn, Transits and Occultations. ArXiv e-prints, January 2010

Chapter 4
Timing the Transit

What then is time? If no one asks me, I know what it is. If I wish to explain it to him who asks, I do not know

Saint Augustine

4.1 Transit Minima

4.1.1 The Exact Solution

I discussed earlier how the minima of S_{P*} occur approximately at the times of inferior and superior conjunction for the transit and occultation respectively. In this section, I will derive the exact solution.

The instants of minima (and maxima) S_{P*} occur when $\mathrm{d}S_{P*}/\mathrm{d}t = 0$. This may be expanded using the chain rule to:

$$\frac{\mathrm{d}S_{P*}}{\mathrm{d}t} = \frac{\mathrm{d}S_{P*}}{\mathrm{d}f_P}\frac{\mathrm{d}f_P}{\mathrm{d}t} \tag{4.1}$$

Since $\dot{f}_P \neq 0$ under any circumstances, then the condition that $\mathrm{d}S_{P*}/dt = 0$ is equivalent to $\mathrm{d}S_{P*}/\mathrm{d}f_P = 0$, which is more manageable mathematically. However, S_{P*} involves a square root function, which I prefer to avoid as it tends to produce more elaborate forms when differentiated relative to functions of integer indices. S_{P*}^2 therefore makes a more useful starting point.

$$\frac{\mathrm{d}(S_{P*}^2)}{\mathrm{d}f_P} = 2S_{P*}\frac{\mathrm{d}S_{P*}}{df_P} \tag{4.2}$$

From the above, one can see that the condition $\mathrm{d}S_{P*}^2/\mathrm{d}f_P = 0$ is equivalent to the condition $\mathrm{d}S_{P*}/\mathrm{d}f_P = 0$, except in the case where $S_{P*} = 0$. However, since

D. M. Kipping, *The Transits of Extrasolar Planets with Moons*, Springer Theses,
DOI: 10.1007/978-3-642-22269-6_4,

$S_{P*} \geq 0$ at all times, then the instant when $S_{P*} = 0$ must correspond to a minimum anyway.

$$S_{P*}^2 = \frac{a_P^2(1-e_P^2)^2}{(1+e_P\cos f_P)^2}\Big[1-\sin^2 i_P \sin^2(\omega_P + f_P)\Big] \tag{4.3}$$

$$\begin{aligned}\frac{\mathrm{d}(S_{P*}^2)}{\mathrm{d}f_P} = -\frac{2a_P^2(1-e_P^2)^2}{(1+e_P\cos f_P)^3}\Big(&(1+e_P\cos f_P)\sin(f_P+\omega_P)\cos(f_P+\omega_P)\sin^2 i_P \\ &- e_P \sin f_P[1-\sin^2 i_P \sin^2(f_P+\omega_P)]\Big)\end{aligned} \tag{4.4}$$

Inspection of $\mathrm{d}S_{P*}^2/\mathrm{d}f_P$ shows a common factor on the outside. This factor has no bearing on the minima and can only reach an extrema if $e_P \geq 1$, which is forbidden for bounded orbits. Therefore, the condition under which minima occur is:

$$\begin{aligned}(1+e_P\cos f_P)\sin(f_P+\omega_P)\cos(f_P+\omega_P)\sin^2 i_P \\ = e_P\sin f_P[1-\sin^2 i_P\sin^2(f_P+\omega_P)]\end{aligned} \tag{4.5}$$

Rearranging and writing in terms of the coefficients of $\cos f_P$, one obtains a quartic equation, of form $R_0 + R_1\cos f_P + R_2\cos^2 f_P + R_3\cos^3 f_P + R_4\cos^4 f_P = 0$, where:

$$R_0 = -4e_P^2(1-\sin^2 i_P\cos^2\omega_P)^2 + \sin^4 i_P\sin^2(2\omega_P) \tag{4.6}$$

$$R_1 = 8e_P\sin^2 i_P(\cos 2\omega_P - \cos^4\omega_P\sin^2 i_P) \tag{4.7}$$

$$R_2 = 4e_P^2 - \sin^2 i_P\left(2+3e_P^2-(2-e_P^2)\cos 2i_P + e_P^2\cos 2\omega_P(3+\cos 2i_P)\right) \tag{4.8}$$

$$R_3 = -2e_P\cos 2\omega_P\sin^2 i_P(3+\cos 2i_P) + 4e_P\sin^4 i_P \tag{4.9}$$

$$R_4 = 4\sin^4 i_P \tag{4.10}$$

The quartic solution has four roots, two of which correspond to the instants where minima occur and two of which correspond to the maxima. The standard analytic solutions for the roots of a quartic equation, known as Ferrari's solution, are unfortunately highly elaborate and take several pages to write out. Further, which root corresponds to a minima or maxima varies depending upon ω_P, e_P and i_P with complex boundary conditions. A final complication comes from the fact I have solved the equation in $\cos f_P$, which means there are twice as many roots (i.e. eight roots) for f_P (concordantly I refer to the problem as a bi-quartic equation).

For these reasons, solving the bi-quartic is clearly possible, but somewhat impractical. One has to find all eight roots and then test each of them to see which one is the true solution. The resulting algorithm is rather inefficient and an alternative approach is clearly desirable.

4.1.2 The Series Expansion Solution

A series expansion approach is attractive as one can continue up to any desired level of precision and yet possess just one root i.e. avoid the root-selection problems encountered with the bi-quartic. The expansion can also be written as a series of much simpler expressions than that encountered in Sect. 4.1.1.

Many methods exist for iterating towards a root with the most commonly used being the Newton-Rhapson method:

$$x_{i+1} = x_i \frac{f(x_i)}{f'(x_i)} \tag{4.11}$$

A good starting point is required for the iteration and so I select the time of inferior conjunction, $f_P = \pi/2 - \omega_P$. I define the function, *f(x)*, to be $\mathrm{d}S_{P*}^2/\mathrm{d}f_P$, as given in (4.4). The final solution for $f_{P,T}$ may be written as:

$$f_{P,T} = \left[\frac{\pi}{2} - \omega_P\right] - \sum_{i=1}^{n} \eta_i^T \tag{4.12}$$

The η_i terms are used to absorb the perturbing terms from the time of inferior conjunction. This notation comes from [1], who used a single η parameter and only presented up to the first-order expansion, whereas here I wish to go up to higher order.

Newton's method produces quite elaborate formulas when used naively. However, one useful trick is to perform a series expansion of x_i using a Taylor series. This Taylor expansion should be performed in a parameter which is very close to zero, so that only a few terms are needed to achieve the required precision. An excellent candidate for this parameter is $\cos^2 i_P$ which is very close to zero for transits. Previous authors, such as [2] and [1], expanded in $\cot^2 i_P$ rather than $\cos^2 i_P$ though, so I will pause to consider the value of such an approach. One wishes to choose the parameter which is closest to zero for $i_P \to \pi/2$. Writing out the first few terms of each one obtains:

$$\cos(\pi/2 - x) = x - \frac{x^3}{6} + \mathcal{O}[x^5]$$
$$\cot(\pi/2 - x) = x + \frac{x^3}{3} + \mathcal{O}[x^5]$$

It can therefore be seen that $\cos^2 i_P$ will approach zero faster than $\cot^2 i_P$. Another subtlety is that if one has performed n iterations with Newton's method, any parts of the series expansion in $\cos^2 i_P$ above order n will change after the next iteration. In other words, they are unstable. Therefore, one must be careful to only Taylor expand up to order n in $\cos^2 i_P$ if one has performed n iterations with Newton's method. Taking into account all of these issues, the first six terms of the η_i^T series are given by:

$$\eta_1^T = +\left(\frac{k_P}{1+h_P}\right)(\cos^2 i_P)^1 \tag{4.13}$$

$$\eta_2^T = +\left(\frac{k_P}{1+h_P}\right)\left(\frac{1}{1+h_P}\right)(\cos^2 i_P)^2 \tag{4.14}$$

$$\eta_3^T = -\left(\frac{k_P}{1+h_P}\right)\left(\frac{-6(1+h_P)+k_P^2(-1+2h_P)}{6(1+h_P)^3}\right)(\cos^2 i_P)^3 \tag{4.15}$$

$$\eta_4^T = -\left(\frac{k_P}{1+h_P}\right)\left(\frac{-2(1+h_P)+k_P^2(-1+3h_P)}{2(1+h_P)^4}\right)(\cos^2 i_P)^4 \tag{4.16}$$

$$\eta_5^T = +\left(\frac{k_P}{1+h_P}\right)\left(\frac{40(1+h_P)^2-40k_P^2(-1+3h_P+4h_P^2)+k_P^4(3-19h_P+8h_P^2)}{40(1+h_P)^6}\right)(\cos^2 i_P)^5 \tag{4.17}$$

$$\eta_6^T = +\left(\frac{k_P}{1+h_P}\right)\left(\frac{24(1+h_P)^2-40k_P^2(-1+4h_P+5h_P^2)+9k_P^4(1-8h_P+5h_P^2)}{24(1+h_P)^7}\right)(\cos^2 i_P)^6 \tag{4.18}$$

where $h_P = e_P \sin\omega_P$ and $k_P = e_p \cos\omega_P$. For the occultation event, one may repeat the process but using the initial guess $3\pi/2 - \omega_P$ instead. The occultation minimum then occurs at:

$$f_{P,O} = \left[\frac{3\pi}{2} - \omega_P\right] - \sum_{i=1}^{n} \eta_i^O \tag{4.19}$$

where:

$$\eta_1^O = -\left(\frac{k_P}{1-h_P}\right)(\cos^2 i_P)^1 \tag{4.20}$$

$$\eta_2^O = -\left(\frac{k_P}{1-h_P}\right)\left(\frac{1}{1-h_P}\right)(\cos^2 i_P)^2 \tag{4.21}$$

$$\eta_3^O = -\left(\frac{k_P}{1-h_P}\right)\left(\frac{6(1-h_P)+k_P^2(1+2h_P)}{6(1-h_P)^3}\right)(\cos^2 i_P)^3 \tag{4.22}$$

$$\eta_4^O = -\left(\frac{k_P}{1-h_P}\right)\left(\frac{2(1-h_P)+k_P^2(1+3h_P)}{2(1-h_P)^4}\right)(\cos^2 i_P)^4 \tag{4.23}$$

$$\eta_5^O = -\left(\frac{k_P}{1-h_P}\right)\left(\frac{40(1-h_P)^2-40k_P^2(-1-3h_P+4h_P^2)+k_P^4(3+19h_P+8h_P^2)}{40(1-h_P)^6}\right)(\cos^2 i_P)^5 \tag{4.24}$$

$$\eta_6^O = -\left(\frac{k_P}{1-h_P}\right)\left(\frac{24(1-h_P)^2 - 40k_P^2(-1-4h_P+5h_P^2) + 9k_P^4(1+8h_P+5h_P^2)}{24(1-h_P)^7}\right)(\cos^2 i_P)^6 \tag{4.25}$$

I was unable to find the series presented in (4.18) and (4.25) in the previous exoplanet, or eclipsing binary, literature above second order and thus conclude they are novel results. The equations have been verified against the solution found from the bi-quartic equation and seem to converge very rapidly. I found that it was not practical to derive the solutions above sixth order. The above equations were computed using *Mathematica* with computation time increasing exponentially with the order number. On a 2 GHz *Intel* processor, the final term requires around 20 mins to generate and I extrapolate that the next order will require two days of continuous processing. However, in practice, the computation uses up all of the available memory before this point and brings the calculation speed down to a much lower pace.

One may compare the η^T-series against the first-order solution presented in [1], who finds:

$$f_{P,T}^{\text{Kopal}} = \frac{\pi}{2} - \omega_P - \left(\frac{k_P}{1+h_P}\right)\cot^2 i_P \tag{4.26}$$

Therefore, provided one acknowledges that $\cot^2 i_P \simeq \cos^2 i_P$ for $i_P \simeq (\pi/2)$, the [1] solution is identical to the first order term in the η^T-series.

It is useful to consider the error formulas of the η-series. If one goes up to order n in the series, the error in $f_{P,T}$ will be $\simeq |\eta_{n+1}^P|$. It is more useful, though, to consider the error in terms of time. To make the conversion from $f \to t$ requires solving Kepler's Equation, which is clearly not desirable due its transcendental nature. However, since the error in f_P will be small, it is justified to approximate $\mathrm{d}f_P \simeq \Delta f_P$ and $\mathrm{d}t \simeq \Delta t$ in (3.11). I will also approximate $f_{P,T} \simeq \pi/2 - \omega_P$, which essentially is saying the speed of the planet at the time of transit minimum is approximately the same as that as at the time of inferior conjunction. This allows one to write that the error on the n th-order η^T-series expansion for the time of transit minimum, τ_T, is[1,2]:

$$(\Delta\tau_T)_n \simeq |\eta_{n+1}^T| \frac{P_P}{2\pi} \frac{(1-e_P^2)^{3/2}}{(1+h_P)^2} \tag{4.27}$$

Similarly, for the occultation minimum:

$$(\Delta\tau_O)_n \simeq |\eta_{n+1}^O| \frac{P_P}{2\pi} \frac{(1-e_P^2)^{3/2}}{(1-h_P)^2} \tag{4.28}$$

---

[1] Note, the notation of using τ for the transit minima is not present in the exoplanet literature, but is adopted in this thesis in an effort to produce a self-consistent notation set spanning both planets and moons (see Appendix A).

[2] Note, in virtually cases in the exoplanet literature this time is known as the mid-transit time, whereas in the eclipsing binary community is known as the eclipse minimum. The second definition is far more accurate as mid-time refers to simply the halfway point of the transit, which is not the same as the minimum (see Appendix A).

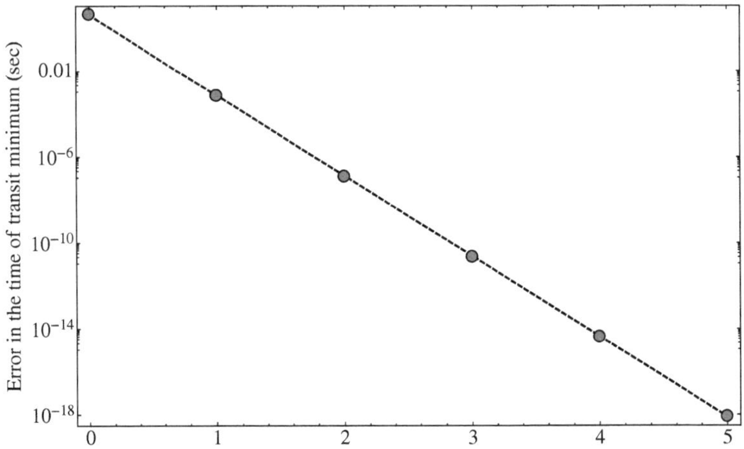

Fig. 4.1 Convergence of the η^T-series for predicting the time of the transit minimum, for the example of the HAT-P-2b system. The very rapid convergence is shown by the best-fit line, indicating a rate of convergence of 8.6 here

I will here illustrate the convergence of the η^T-series with an example of a known transiting system. To maximize the number of iterations needed, I require a short-period highly eccentric planet. HAT-P-2b [3] gives an ideal test with $P_P = 5.6334729$d, $e_P = 0.5171$, $\omega_P = 185.22°$ and $i_P = 86.72°$. In Fig. 4.1, I show the resulting errors in τ_T as a function of the order used from the η^T series. Using a zeroth-order expansion gives an error of 4.5 s, which is generally too large, but going to first-order reduces this to less than a millisecond as a result of the very rapid convergence. Although the general speed of convergence is variable, this typical example has a convergence of rate of 8.6, illustrating how rapidly the series works.

4.2 Transit Duration for a Circular Orbit

4.2.1 Fundamental Equations

Calculating the duration of a transit is non-trivial and I begin the discussion by considering the simple case of a circular orbit. The first point to establish is that numerous definitions for the transit duration exist in the literature. As was seen earlier in Sect. 3.3.3 and Fig. 3.3, there are four principal contact points which define the transit, t_I, t_{II}, t_{III} and t_{IV}.t_I and t_{IV} occur when $S_{P*} = (1 + p)$ and t_{II} and t_{III} occur when $S_{P*} = (1 - p)$. I define the time between any two contact points as T_{xy}, where x and y are the contact points.[3] T_{14} is sometimes referred to as the

[3] Note that x and y are commutable.

"total duration" and T_{23} as the "flat-bottomed duration" or confusingly the "full duration". T_{12} and T_{34} tend to have a negligible difference (even for very eccentric systems) and are therefore often grouped together and called the "ingress/egress duration".

The duration between any two instances is found by computing the time between the planetary true anomalies at those points. For contact points x and y, I define the associated true anomalies as $f_{P,x}$ and $f_{P,y}$ and thus the time between these points can be found by integrating (3.11):

$$\begin{aligned} T_{xy} &= \int_{f_P=f_{P,x}}^{f_{P,y}} \frac{\mathrm{d}t}{\mathrm{d}f_P} \mathrm{d}f_P \\ &= \frac{P_P}{2\pi} \int_{f_P=f_{P,x}}^{f_{P,y}} \frac{(1-e_P^2)^{3/2}}{(1+e_P\cos f_P)^2} \mathrm{d}f_P \end{aligned} \tag{4.29}$$

Recall earlier in Sect. 3.1.2, (3.7), that the mean anomaly is defined to scale linearly with time. Therefore, by virtue of its very definition, the mean anomaly is also directly related to the transit duration:

$$\begin{aligned} T_{xy} &= \frac{P_P}{2\pi}(\mathfrak{M}_{P,y} - \mathfrak{M}_{P,x}) \\ \Rightarrow \Delta\mathfrak{M}_P &= \int_{f_P=f_{P,x}}^{f_{P,y}} \frac{(1-e_P^2)^{3/2}}{(1+e_P\cos f_P)^2} \mathrm{d}f_P \end{aligned} \tag{4.30}$$

From the above, it can therefore be seen that $\Delta\mathfrak{M}_P = \Delta f_P$ if $e_P = 0$ i.e. a circular orbit. In such a case, the transit duration may therefore be simply expressed as:

$$T_{xy}^{\mathrm{circ}} = \frac{P_P}{2\pi}(f_{P,y} - f_{P,x}) \tag{4.31}$$

The outstanding problem is now to solve for $f_{P,x}$ and $f_{P,y}$. Let us assume that contact points x and y have the same sky-projected planet-star separation i.e. $S_{P*}(f_{P,x}) = S_{P*}(f_{P,y})$. Consequently, the true anomalies $f_{P,x}$ and $f_{P,y}$ may be found by solving the following for f_P:

$$S_{P*} = (a_P/R_*)\left(\frac{1-e_P^2}{1+e_P\cos f_P}\right)\sqrt{1-\sin^2 i_P \sin^2(\omega_P + f_P)} \tag{4.32}$$

For a circular orbit, the e_P terms vanish, but an ω_P term would persist in the function $\sin^2(\omega_P + f_P)$. For a circular orbit, the transit minimum occurs exactly at the time of inferior conjunction (visible by setting $e_P = 0$ in the η^T-series in (4.18)). I therefore define the difference in the true anomaly between the instant of inferior conjunction and the relevant contact points as $\breve{f}_P$. Therefore, $\breve{f}_{P,x} = \pi/2 - \omega_P - f_{P,x}$ and $\breve{f}_{P,y} = \pi/2 - \omega_P - f_{P,y}$ and consequently $\Delta \breve{f}_P = \Delta f_P$. Equation (4.32) now becomes:

$$S_{P*} = (a_P/R_*)\sqrt{1 - \sin^2 i_P \cos^2 \breve{f}_P^{\mathrm{circ}}} \tag{4.33}$$

Solving for $\breve{f}_P$ yields:

$$\breve{f}_{P,x}^{\mathrm{circ}} = -\arcsin\left[\sqrt{\frac{S_{P*}^2 - b_{P,T}^2}{(a_P/R_*)^2 - b_{P,T}^2}}\right]$$

$$\breve{f}_{P,y}^{\mathrm{circ}} = \arcsin\left[\sqrt{\frac{S_{P*}^2 - b_{P,T}^2}{(a_P/R_*)^2 - b_{P,T}^2}}\right]$$

$$\Delta f_P^{\mathrm{circ}} = \Delta \breve{f}_P^{\mathrm{circ}} = 2\arcsin\left[\sqrt{\frac{S_{P*}^2 - b_{P,T}^2}{(a_P/R_*)^2 - b_{P,T}^2}}\right] \tag{4.34}$$

Plugging this into (4.31) gives an exact solution for the duration of a transit, for circular orbits.

$$T_{xy}^{\mathrm{circ}} = \frac{P_P}{\pi}\arcsin\left[\sqrt{\frac{S_{P*}^2 - b_{P,T}^2}{(a_P/R_*)^2 - b_{P,T}^2}}\right] \tag{4.35}$$

Equation (4.35) is only valid if contact points x and y have the same S_{P*}. Two cases which are therefore applicable are T_{14} and T_{23}:

$$T_{14}^{\mathrm{circ}} = \frac{P_P}{\pi}\arcsin\left[\sqrt{\frac{(1+p)^2 - b_{P,T}^2}{(a_P/R_*)^2 - b_{P,T}^2}}\right] \tag{4.36}$$

$$T_{23}^{\mathrm{circ}} = \frac{P_P}{\pi}\arcsin\left[\sqrt{\frac{(1-p)^2 - b_{P,T}^2}{(a_P/R_*)^2 - b_{P,T}^2}}\right] \tag{4.37}$$

For the ingress/egress duration, one cannot use (4.35) since the contact points now have different values for the sky-projected separation. However, one may retrace the steps in the above derivation to obtain such a result and find that the time difference between contact points 1&2 and 3&4 give an identical result:

$$T_{12}^{\text{circ}} = T_{34}^{\text{circ}} = \frac{P_P}{2\pi}\left(\arcsin\left[\sqrt{\frac{(1+p)^2 - b_{P,T}^2}{(a_P/R_*)^2 - b_{P,T}^2}}\right] - \arcsin\left[\sqrt{\frac{(1-p)^2 - b_{P,T}^2}{(a_P/R_*)^2 - b_{P,T}^2}}\right]\right)$$

$$T_{12}^{\text{circ}} = T_{34}^{\text{circ}} = \frac{T_{14}^{\text{circ}} - T_{34}^{\text{circ}}}{2} \tag{4.38}$$

where on the last line I have shown how the ingress/egress duration may be written as a linear combination of the two durations previously defined.

4.2.2 Parameter Retrieval

The parameters p, $b_{P,T}$ and (a_P/R_*) should now be clearly identifiable as the main players in these equations. These three parameters combined with the time of the transit minimum, τ_T, provide the four-observables of transit light curves for both circular and eccentric orbits. It has already been seen how the ingress/egress duration is simply a linear combination of T_{14} and T_{23} and thus it does not provide any new information. In other words, one only needs two quantities to define the transit durations. Since the transit depth gives p, and the timing gives τ_T, the outstanding challenge one faces is whether these two durations can be converted into $b_{P,T}$ and (a_P/R_*).

Seager and Mallén-Ornelas [4] were the first to appreciate this point and successfully provide a solution. Through rearranging the expressions for T_{14}^{circ} and T_{23}^{circ}, they showed:

$$b_{P,T}^{\text{circ}} = \left[\frac{(1-p)^2 - \frac{\sin^2(T_{23}^{\text{circ}}\pi/P_P)}{\sin^2(T_{14}^{\text{circ}}\pi/P_P)}(1+p)^2}{1 - \frac{\sin^2(T_{23}^{\text{circ}}\pi/P_P)}{\sin^2(T_{14}^{\text{circ}}\pi/P_P)}}\right]^{1/2} \tag{4.39}$$

$$\left(\frac{a_P}{R_*}\right)^{\text{circ}} = \left[\frac{(1+p)^2 - [b_{P,T}^{\text{circ}}]^2(1 - \sin^2(T_{14}^{\text{circ}}\pi/P_P)}{\sin^2(T_{23}^{\text{circ}}\pi/P_P)}\right]^{1/2} \tag{4.40}$$

If the semi-major axis of the orbit, a_P, and the period, P_P, was known (through timing of consecutive transits) then it would be possible to determine the mass of the star (or technically $M_* + M_P$) through Kepler's Third Law. However, nature throws a spanner into the works because one cannot determine a_P, only (a_P/R_*). As a result, instead of being able to measure M_*, one obtains the average stellar density, ρ_*:

$$\frac{a_P^3}{G(M_* + M_P)} = \left(\frac{P_P}{2\pi}\right)^2$$
$$\frac{a_P^3}{R_*^3} = \left(\frac{P_P}{2\pi}\right)^2 G \left(\frac{M_*}{R_*^3} + \frac{M_P}{R_*^3}\right)$$
$$\rho_* + p^3 \rho_P = \frac{3\pi (a_P/R_*)^3}{G P_P^2} \tag{4.41}$$

The importance of (4.41) cannot be overstated. This pioneering insight by Seager and Mallén-Ornelas [4] now forms the standard method of determining the properties of a transiting planet's host star. A typical approach is to compute stellar evolution isochrones of a main sequence star (e.g. [5, 6]) and see which isochrones best-match the measured stellar density, as well as other stellar parameters determined spectroscopically (e.g. effective temperature of the star, T_{eff}, metallicity, [Fe/H], and surface gravity, $\log g$).

Of course, the suggested method is actually somewhat restricted in that the equations of Seager and Mallén-Ornelas [4] are only valid for circular orbits. A large fraction of the known transiting planets have been found to maintain non-zero eccentricities and thus this assumption is critically flawed. Clearly, what is required is an updated version of the Seager and Mallén-Ornelas [4] equations but accounting for eccentricity. To do this, one first needs to derive the transit duration for such systems, which will be provided in Sect. 4.3. First though, I will introduce an additional definition of the transit duration, which will be very useful in the rest of this thesis.

4.2.3 The $\tilde{T}$ Duration

So far, I have considered the durations between the four contact points defined in Sect. 3.3.3 and Fig. 3.3. However, Carter et al. [7] proposed an additional, and very useful, definition of the transit duration which does not involve these four points. $\tilde{T}$ is defined as the time it takes for the planet's centre to cross the stellar limb and exit under the same condition.[4] Mathematically, these points occur when $S_{P*} = 1$. Critically, this definition excludes the parameter p and can be expressed in terms of $b_{P,T}$ and (a_P/R_*) only (for a circular orbit). Naturally, if one has a measurement error on all three observables, then these will propagate into the estimates of the various durations, but $\tilde{T}$ is only exposed to two of these errors and thus tends to be found with the highest signal-to-noise (SNR). For this reason, $\tilde{T}$ is very useful for TDV (transit duration variation) studies.

On a brief aside, [7] showed that $\tilde{T}$ has the highest SNR for a trapezoid-approximated light curve and thus limb darkening is notably absent. If one uses the same bandpass for a set of TDV measurements, one may safely fix the quadratic

[4] Since $\tilde{T}$ is not defined in terms of the four contact points, it is not possible to use the T_{xy} notation employed earlier, and thus the tilde symbol is used instead.

limb darkening coefficients to some reasonable estimate and $\tilde{T}$ will tend to give the lowest errors. However, if one compares multiple wavelengths, it must be acknowledged that the limb darkening will vary across each light curve and thus should be fitted for. This tends to increase the error on $\tilde{T}$ in particular and, often, it is no longer the best duration parameter. However, performing TDV across multiple wavelengths should be avoided anyway, based upon the detection criterion C5. For the Seager and Mallén-Ornelas [4] (SMO03) model, it is trivial to show that $\tilde{T}$ is given by using $S_{P*} = 1$ in (4.35):

$$\tilde{T}^{\mathrm{circ}} = \frac{P_P}{\pi} \arcsin\left[\sqrt{\frac{1 - b_{P,T}^2}{(a_P/R_*)^2 - b_{P,T}^2}}\right] \tag{4.42}$$

It should be stressed at this point that $\tilde{T}^{\mathrm{circ}} \neq (T_{14}^{\mathrm{circ}} + T_{23}^{\mathrm{circ}})/2$.

4.3 Transit Duration for an Eccentric Orbit

4.3.1 The Exact Solution

The duration between any two true anomalies is most concisely expressed in terms of the change in the mean anomaly between those points, as first shown in (4.30).

$$\begin{aligned} T_{xy} &= \frac{P_P}{2\pi} \Delta \mathfrak{M}_P \\ \Delta \mathfrak{M}_P &= \int_{f_P = f_{P,x}}^{f_{P,y}} \frac{(1 - e_P^2)^{3/2}}{(1 + e_P \cos f_P)^2} \mathrm{d} f_P \end{aligned} \tag{4.43}$$

For circular orbits, $\Delta \mathfrak{M}_P = \Delta f_P$, but this is not the case for eccentric ones. Integrating the above equation yields:

$$\begin{aligned} \Delta \mathfrak{M}_P = & \left(2(1 - e_P^2) \arctan\left[\sqrt{\frac{1 - e_P}{1 + e_P}} \tan \frac{f_{P,y}}{2}\right] - \frac{e_P (1 - e_P^2)^{3/2} \sin f_{P,y}}{1 + e_P \cos f_{P,y}} \right) \\ & - \left(2(1 - e_P^2) \arctan\left[\sqrt{\frac{1 - e_P}{1 + e_P}} \tan \frac{f_{P,x}}{2}\right] - \frac{e_P (1 - e_P^2)^{3/2} \sin f_{P,x}}{1 + e_P \cos f_{P,x}} \right) \end{aligned} \tag{4.44}$$

Now, the only remaining challenge is to find $f_{P,x}$ and $f_{P,y}$. As was seen before, for T_{14}, one has:

$$S_{P*} = (a_P/R_*) \left(\frac{1 - e_P^2}{1 + e_P \cos f_P} \right) \sqrt{1 - \sin^2 i_P \sin^2(\omega_P + f_P)} \tag{4.45}$$

Solving the above for f_P is a challenging problem. As was done in Sect. 4.1, the problem is easier if one changes all f_P terms to $\cos f_P$ parameters. This, again, yields a quartic equation $Q_0 + Q_1 \cos f_P + Q_2 \cos^2 f_P + Q_3 \cos^3 f_P + Q_4 \cos^4 f_P$, where:

$$Q_0 = \left(\csc^2 i_P (S_{P*}^2 - \Lambda_P^2) + \Lambda_P^2 \cos^2 \omega_P\right)^2 \tag{4.46}$$

$$Q_1 = 2 e_P S_{P*}^2 \csc^2 i_P \left(2 \csc^2 i_P (S_{P*}^2 - \Lambda_P^2) + \Lambda_P^2 + \Lambda_P^2 \cos 2\omega_P\right) \tag{4.47}$$

$$\begin{aligned} Q_2 = {} & \Lambda_P^2 \cos 2\omega_P \left(S_{P*}^2 \csc^2 i_P (-2 + e_P^2) - \Lambda_P^2 + 2\Lambda_P^2 \csc^2 i_P\right) \\ & - \Lambda_P^4 + e_P^2 S_{P*}^2 \Lambda_P^2 \csc^2 i_P + 2 \csc^4 i_P e_P^2 S_{P*}^2 (3 S_{P*}^2 - \Lambda_P^2) \end{aligned} \tag{4.48}$$

$$Q_3 = 4 e_P S_{P*}^2 \csc^2 i_P (e_P^2 S_{P*}^2 \csc^2 i_P - \Lambda_P^2 \cos 2\omega_P) \tag{4.49}$$

$$Q_4 = \Lambda_P^4 - 2 e_P^2 S_{P*}^2 \Lambda_P^2 \cos 2\omega_P \csc^2 i_P + e_P^4 S_{P*}^4 \csc^4 i_P \tag{4.50}$$

$$\Lambda_P = (a_P/R_*)(1 - e_P^2) \tag{4.51}$$

The quartic nature of the solution for this problem was first identified by Kipping [8]. As the solution is presented in $\cos f_P$ rather than f_P, (4.51) actually has eight roots in the same way as found for the transit minima, and thus is again a bi-quartic equation. The large number of solutions causes similar problems as before and a root selection routine is required. As before, I will seek methods to avoid the bi-quartic.

4.3.2 Approximate Solutions

One way to avoid the bi-quartic is to make an approximation. A useful approximation which can be made is that $\varrho_P(f_P) \simeq \varrho_{P,T} = \varrho_P(f_P = f_{P,T})$, i.e. the planet-star separation is approximately a constant value given by the planet-star separation at the transit minimum. I will also make the approximation $f_{P,T} \simeq (\pi/2 - \omega_P)$ to simplify the mathematics, which can be justified at the level of a few seconds accuracy for even the most extreme systems (e.g. Fig. 4.1).

Modifying the definition of the transit impact parameter to account for the altered planet-star separation so that $b_{P,T} = (a_P/R_*)\varrho_{P,T} \cos i_P$, the true anomalies satisfy (4.52), which is subsequently solved as was done before for the circular orbits (Sect. 4.2.1):

$$\begin{aligned} S_{P*} &\simeq (a_P/R_*)\varrho_{P,T} \sqrt{1 - \sin^2 i_P \cos^2 \breve{f}_P} \\ \Delta f_P = \Delta \breve{f}_P &\simeq 2 \arcsin\left[\sqrt{\frac{S_{P*}^2 - b_{P,T}^2}{(a_P/R_*)^2 \varrho_{P,T}^2 - b_{P,T}^2}}\right] \end{aligned} \tag{4.52}$$

Having assigned the true anomaly at the moment of the transit minimum and the change in the true anomaly over the transit event, the necessary tools to compute the duration have been found. Unlike the circular orbit case, there are actually two options as to how to proceed here. Firstly, one could use the solutions for $f_{P,x}$ and $f_{P,y}$ and feed them into the (4.44) to get $\Delta\mathfrak{M}_P$ and then use this in (4.29) to get the duration. Given that (4.44) has two principal functions, this will yield a "two-term" expression for the duration. Alternatively, one could assume $\Delta\mathfrak{M}_P \simeq \Delta f_P$ and avoid (4.44) altogether, which would yield a "one-term" expression. I will investigate both approaches here, followed by alternative methods adopted by some previous authors.

4.3.2.1 Two-Term Expression

By combining (4.44) with (4.29), one may obtain a final expression for the duration, which I label T^{two} ("two-term"):

$$T_{xy}^{\text{two}} = \frac{P_P}{2\pi}\Bigg[\arctan\left[\frac{\cos(\Delta f_P/2) + e_P\cos(\langle f_P\rangle)}{\sqrt{1-e_P^2}\sin(\Delta f_P/2)}\right] + \frac{e_P(1-e_P^2)^{1/2}\sin(\Delta f_P/2)(e_P\cos(\Delta f_P/2)+\cos(\langle f_P\rangle))}{(1-e_P^2)\sin^2(\langle f_P\rangle)+(e_P\cos(\Delta f_P/2)+\cos(\langle f_P\rangle))^2}\Bigg] \qquad (4.53)$$

where I have used Δf_P from (4.52) and defined $\langle f_P\rangle = (f_{P,x} + f_{P,y})/2$. The approximate entry for this latter term would be $\langle f_P\rangle \simeq \pi/2 - \omega_P$. Testing T_{xy}^{two} for the exact solutions for $f_{P,T}$ and Δf_P provided precisely the correct transit duration for all e_P, as expected. However, I found that using approximate entries for these terms severely limited the precision of the derived equation for large e_P (the results of numerical tests will be discussed later in Sect. 4.3.3).

The source of the problem is visible in (4.53), which consists of taking the difference between two terms. Both terms are of comparable magnitude for large e_P and thus one is obtaining a small term by taking the difference between two large terms. These kinds of expressions are very sensitive to slight errors. In this case, the error is from using approximate entries for $\langle f_P\rangle$ and in particular Δf_P. Next, I will consider a possible "one-term" expression, which avoids the problem of taking the difference between two comparable-magnitude terms and thus produces greater numerical stability.

4.3.2.2 One-Term Expression

There are numerous possible methods for finding "one-term" transit duration expressions. The first one I consider is to assume $\Delta\mathfrak{M}_P \simeq \Delta f_P$ thereby avoiding the two-term nature of (4.44):

$$
\begin{aligned}
T_{xy}^{\text{one}} &= \frac{P_P}{2\pi} \frac{\varrho_{P,T}^2}{\sqrt{1-e_P^2}} \Delta f_P \\
&= \frac{P_P}{\pi} \frac{\varrho_{P,T}^2}{\sqrt{1-e_P^2}} \arcsin\left[\sqrt{\frac{S_{P*}^2 - b_{P,T}^2}{(a_P/R_*)^2 \varrho_{P,T}^2 - b_{P,T}^2}}\right]
\end{aligned} \tag{4.54}
$$

where (4.52) has been used for Δf_P. Another derivation would be to assume the planet takes a tangential orbital velocity and constant orbital separation from the planet, sweeping out an arc of length $r_{P,T}\Delta f_P$. It is trivial to show that this argument will lead to precisely the same expression for T_{xy}^{one}.

4.3.2.3 Tingley and Sackett (2005) Equation

I will also describe the derivations of two other commonly used formulas in the exoplanet literature. The Seager and Mallén-Ornelas [4] (SMO03) formula has already been discussed in Sect. 4.2 and so will not be covered again.

The first alternative I discuss is that of Tingley and Sackett [9](TS05), which has been used by numerous authors since (e.g. [10, 11]). It is also forms the basis of a light curve parameter fitting set proposed by Bakos et al. [3]. There are two critical assumptions made in the derivation of the TS05 formula. The first of these is that:

- The planet-star separation, r_P, is constant during the planetary transit event and equals $r_{P,T}$

This is the same assumption made in the derivation of the T_{xy}^{one} equation. Under this assumption, TS05 quote the following expression for T_{xy}^{TS05} (changing to consistent notation).

$$
T_{xy}^{\text{TS05}} = \frac{r_{P,T}\Delta\phi}{v_{P,T}} \tag{4.55}
$$

where TS05 define $v_{P,T}$ as the planet's orbital velocity at the transit minimum and $\Delta\phi$ as "the eccentric angle" between the contact points. In the standard notation, there is no such parameter defined strictly as the "eccentric angle" and thus I initially assumed that TS05 were referring to the eccentric anomaly. However, substituting the relevant terms for $r_{P,T}$ and $v_{P,T}$ gives:

$$
T_{xy}^{\text{TS05}} = \frac{P_P}{2\pi} \frac{\varrho_{P,T}^2}{\sqrt{1-e_P^2}} \Delta\phi \tag{4.56}
$$

By comparing (4.56) to (4.54), it is clear that $\Delta\phi = \Delta f_P$ (also note that (4.54) was derived under precisely the same assumptions as that assumed by TS05 at this stage of the derivation). I therefore conclude that the term TS05 refer to as the "eccentric angle" in fact refers to the true anomaly. This is an important point to make because the derivation of the TS05 equation would otherwise be very difficult to understand by those working outside of the field. Continuing the derivation from this point, the second assumption made by TS05 is:

- The planet-star separation is much greater than the stellar radius, $r_{P,T} \gg R_*$

Critically, this assumption was not made in the derivation of T^{one} or T^{two}. Using this assumption, TS05 propose that (replacing $\Delta\phi \to \Delta f_P$ and $1 + p$ to the more general form of S_{P*}, to remain consistent with the notations used in this thesis):

$$\Delta f_P^{\text{TS05}} = \arcsin\left[2\frac{\sqrt{S_{P*}^2 - (a_P/R_*)^2\varrho_{P,T}^2\cos^2 i_P}}{(a_P/R_*)\varrho_{P,T}}\right] \tag{4.57}$$

$$\Delta f_P^{\text{TS05}} \simeq 2\frac{\sqrt{S_{P*}^2 - (a_P/R_*)^2\varrho_{P,T}^2\cos^2 i_P}}{(a_P/R_*)\varrho_{P,T}} \tag{4.58}$$

where TS05 use (4.58) rather than (4.57) in the final version of T^{TS05}. Therefore, TS05 effectively make a small-angle approximation for Δf_P, which is a consequence of assuming $r_{P,T} \gg R_*$. I argue here that losing the arcsin function does not offer any great simplification of the transit duration equation but does lead to an unnecessary source of error in the resultant expression, in particular for close-in orbits, which is common for transits. Also note that even (4.57) exhibits differences to (4.52).

Firstly, inside the arcsin function, the factor of $\csc i_P$ is missing, which is present in both the derivation presented in (4.54) and the derivation of SMO03 for circular orbits, (4.35). The absence of this term can be understood as a result of the $r_{P,T} \gg R_*$ assumption. As $r_{P,T} \to \infty$, in order to maintain a transit event, one must have $i_P \to (\pi/2)$.

Secondly, the expression I presented for Δf_P earlier in (4.52) has the factor of two present outside of the arcsin function, whereas TS05 have this factor inside the function. Further more, the SMO03 derivation also predicts that the factor of two should be outside of the arcsin function and this expression is known to be an exact solution for circular orbits. In a small angle approximation, $\arcsin 2x \simeq 2 \arcsin x$, but moving the factor of 2 to within the arcsin function seems to serve no purpose except to invite further error into the expression. As a result of these differences, the T_{xy}^{TS05} expression does not reduce down to the original SMO03 equation and is given by:

$$T_{xy}^{\text{TS05}} = \frac{P_P}{\pi}\frac{\varrho_{P,T}}{\sqrt{1-e_P^2}}\frac{\sqrt{S_{P*}^2 - (a_P/R_*)^2\varrho_{P,T}^2\cos^2 i_P}}{(a_P/R_*)} \tag{4.59}$$

4.3.2.4 Winn (2010) Equation

Winn [12] (W10) proposed an expression for T_{xy} based upon modification to the SMO03 equation. The first change was to modify the impact parameter from $(a_P/R_*)\cos i_P \rightarrow \varrho_{P,T}(a_P/R_*)\cos i_P$, i.e. to allow for the altered planet-star separation for eccentric orbits, as was done in Sect. 4.3.2. Secondly, the altered planetary velocity should also be incorporated. W10 propose that a reasonable approximation for the transit duration is obtained by multiplying the SMO03 expressions by the following ratio:

$$\frac{\frac{\mathrm{d}X_P}{\mathrm{d}t}(f_{P,T})[e_P = 0]}{\frac{\mathrm{d}X_P}{\mathrm{d}t}(f_{P,T})} = \frac{\varrho_{P,T}}{\sqrt{1-e_P^2}} \tag{4.60}$$

where X_P is given in (3.6). This yields a new transit duration equation of:

$$T_{xy}^{\mathrm{W10}} = \frac{P_P}{\pi}\frac{\varrho_{P,T}}{\sqrt{1-e_P^2}} \arcsin\left[\frac{1}{(a_P/R_*)}\left(\frac{S_{P*}^2 - (a_P/R_*)^2\varrho_{P,T}^2\cos^2 i_P}{\sin^2 i_P}\right)^{1/2}\right] \tag{4.61}$$

Firstly, note an obvious improvement of the W10 expression is that one recovers the original SMO03 equation for $e_P = 0$. Secondly, comparison to the equation for T_{xy}^{one} (4.54) reveals that the two expressions are very similar except for the position of an extra $\varrho_{P,T}$ term. Indeed, the T_{xy}^{one} and T_{xy}^{W10} expressions are equivalent in the small-angle approximation.

4.3.3 Numerical Testing of the Approximations

4.3.3.1 Example Systems

Insights into the robustness and accuracy of the various expressions may be obtained through numerical testing of the various approximate expressions. I here compare the accuracy of the four expressions: $\tilde{T}^{\mathrm{TS05}}$, $\tilde{T}^{\mathrm{W10}}$, $\tilde{T}^{\mathrm{one}}$ and $\tilde{T}^{\mathrm{two}}$. These expressions depend only on five parameters[5]: P_P, (a_P/R_*), $b_{P,T}$, $e_P \sin\omega_P$ and $e_P\cos\omega_P$. One of the clearest ways of comparing the equations is to consider a typical transiting exoplanet example with system parameters for (a_P/R_*) and $b_{P,T}$, and vary the eccentricity parameters. P_P may be selected by simply assuming a star of Solar density and using (4.41).

I created a 1,000 by 1,000 grid of $e_P \sin\omega_P$ and $e_P\cos\omega_P$ values from −1 to 1 in equal steps. Grid positions for hyperbolic orbits ($e_P > 1$) are excluded. I then

[5] There is no dependence on p because I am using the $\tilde{T}$ definition of the duration to deliberately avoid p.

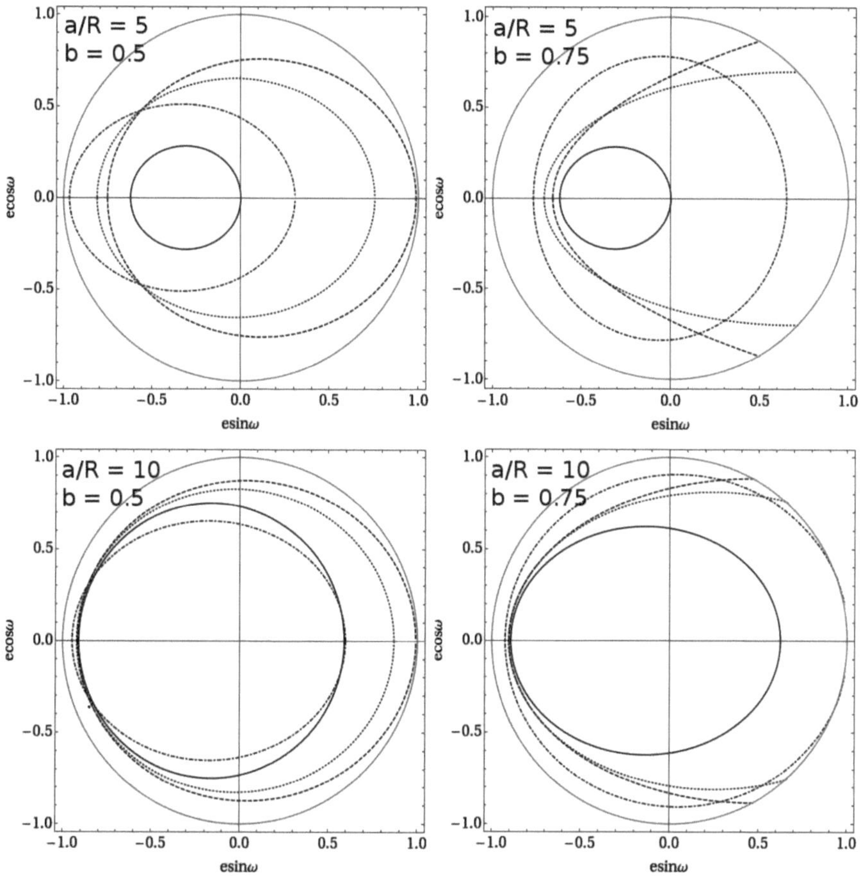

Fig. 4.2 Loci of points for which the accuracy is better than 99% for all four candidate expressions, as a function of eccentricity. The $\tilde{T}^{\text{one}}$ expression offers both consistency and excellent accuracy. Other system parameters fixed to typical transit values. Black solid is for $\tilde{T}^{\text{TS05}}$, dotted is for $\tilde{T}^{\text{W10}}$, dashed is for $\tilde{T}^{\text{one}}$ and dot-dashed for $\tilde{T}^{\text{two}}$. The gray ellipse represents the allowed physical limits. Parameter subscripts removed for clarity

calculated the transit duration through the exact solution of the bi-quartic equation, $\tilde{T}^{\text{biquartic}}$, plus all four approximate formulas. Then, the fractional deviation of each equation from the true solution is found using:

$$\mathcal{D}^{\text{candidate}} = \frac{\tilde{T}^{\text{candidate}} - \tilde{T}^{\text{biquartic}}}{\tilde{T}^{\text{biquartic}}} \tag{4.62}$$

The loci of points for which the deviation is less than 1% (i.e. $\mathcal{D}^{\text{candidate}} < 0.01$) may then be computed. In Fig. 4.2, I present four such plots for different choices of (a_P/R_*) and $b_{P,T}$. The plot reveals several interesting features:

- T^{one} consistently yields the largest loci
- T^{two} is sometimes accurate and sometimes not, demonstrating that the approximation is not stable
- T^{W10} also yields consistently large loci
- T^{TS05} consistently yields the smallest loci

4.3.3.2 Additional Tests

I also briefly discuss additional tests which I performed for two sets of 10^7 different hypothetical transiting exoplanet systems; one for eccentricities $0.0 < e_P < 1$ and the other for $0.9 < e_P < 1$. In all cases, I randomly generated[6] the system parameters weighted by the transit probability and calculated the deviation of the various formulas.

I found that the $\tilde{T}^{\text{one}}$ expression was consistently the most accurate, with the W10 of similar accuracy but higher asymmetry. I therefore find that the results yield a overall preference for the $\tilde{T}^{\text{one}}$ approximation. Note that authors using the T^{W10} formulation can also expect an extremely good approximation but I will only consider using T^{one} for the later derivations. I define the "improvement" of the $\tilde{T}^{\text{one}}$ expression relative to the $\tilde{T}^{\text{TS05}}$ equation (since this is the previously most cited expression) as:

$$\mathcal{I}^{\text{one}} = [(\mathcal{D}^{\text{TS05}}/\mathcal{D}^{\text{one}}) - 1] * 100 \tag{4.63}$$

where $\mathcal{I}$ is measured in %. One can see that if TS05 gives a lower deviation (i.e. more accurate solution), one will obtain $\mathcal{I} < 0\%$ whereas if the candidate expression gives a closer solution one obtains $\mathcal{I} > 0\%$ and is essentially the percentage improvement in accuracy. For the range $0 < e_P < 1$, I find that the mean value of this parameter is $\mathcal{I}^{\text{one}} = 210\%$ and for the range $0.9 < e_P < 1$ I find $\mathcal{I}^{\text{one}} = 458\%$. Note that one caveat of these tests is that they are sensitive to the a-priori inputs. In conclusion, the qualitative, rather than quantitative, aspects of these results offer the greatest insights.

For the case of spaced-based photometry, the typical measurement uncertainty on $\tilde{T}$ will be $\sim 0.1\%$ in most cases [13]. I find, on average, that $\tilde{T}^{\text{one}}$ is accurate to 0.1% or better over a range of $|e_P \sin \omega_P| < 0.5$ and $|e_P \cos \omega_P| < 0.85$.

4.3.4 The Series Expansion Solution

I also investigated whether a series expansion solution could be obtained in a similar way as was done for the transit minima in Sect. 4.1.2. The equation which needs to be solved may be written as:

---

6 With uniform probability distributions.

$$0 = \frac{(a_P/R_*)^2(1-e_P^2)^2}{(1+e_P\cos f_P)^2}[1-\sin^2 i_P \sin^2(\omega_P + f_P)] - S_{P*}^2 \tag{4.64}$$

I tried using Newton's method again with initial starting points for $f_{P,x}$ and $f_{P,y}$ being $\pi/2 - \omega_P - \Delta f_P/2$ and $\pi/2 - \omega_P + \Delta f_P/2$, where I define Δf_P using (4.52). The resulting expressions do not yield elegant forms despite efforts to perform Taylor expansions in numerous terms. Further, computational memory limits restrict the calculation in *Mathematica* to just first order. As a result of the limited use of these expressions, I do not present them here.

4.3.5 The Consequences of Using Circular Expressions for Eccentric Orbits

SMO03 showed that the $1^{\text{st}}-4^{\text{th}}$ contact duration, T_{14}, and the 2nd the 3^{rd}contact duration, T_{23}, may be used to derive (a_P/R_*) and $b_{P,T}$, and consequently the derivative terms such as i_P and ρ_*. I here consider how biased these retrieved parameters would be if one used the circular equations for an eccentric orbit. Such a circumstance may occur for newly discovered transiting planets with only a few RV points and thus a simple circular model is erroneously adopted for the sake of simplicity. From here, I will employ the T^{one} expression for the transit duration, as this equation has been shown to provide the greatest accuracy in Sect. 4.3.3. The transit durations T_{14}^{circ} and T_{23}^{circ} are given (4.37), but for the T_{xy}^{one} solutions I utilize (4.54) to get:

$$T_{14}^{\text{one}} = \frac{P_P}{\pi}\frac{\varrho_{P,T}^2}{\sqrt{1-e_P^2}} \arcsin\left(\frac{\sqrt{(1+p)^2 - (a_P/R_*)^2\varrho_{P,T}^2\cos^2 i_P}}{(a_P/R_*)\varrho_{P,T}\sin i_P}\right) \tag{4.65}$$

$$T_{23}^{\text{one}} = \frac{P_P}{\pi}\frac{\varrho_{P,T}^2}{\sqrt{1-e_P^2}} \arcsin\left(\frac{\sqrt{(1-p)^2 - (a_P/R_*)^2\varrho_{P,T}^2\cos^2 i_P}}{(a_P/R_*)\varrho_{P,T}\sin i_P}\right) \tag{4.66}$$

Using (4.37), SMO03 show that the impact parameter of a transiting planet on a circular orbit may be determined by using:

$$[b_{P,T}^{\text{circ}}]^2 = \frac{(1-p)^2 - \frac{\sin^2(T_{14}^{\text{circ}}\pi/P_P)}{\sin^2(T_{14}^{\text{circ}}\pi/P_P)}(1+p)^2}{1 - \frac{\sin^2(T_{23}^{\text{circ}}\pi/P)}{\sin^2(T_{14}^{\text{circ}}\pi/P_P)}} \tag{4.67}$$

Now consider that an observer is unaware that the orbit is eccentric and makes the simple, but false, assumption of a circular orbit. In this case, the observer would assign $T_{xy}^{\text{circ}} = T_{xy}^{\text{biquartic}} \simeq T_{xy}^{\text{one}}$. As a result of this assumption, the observer would falsely calculate an impact parameter:

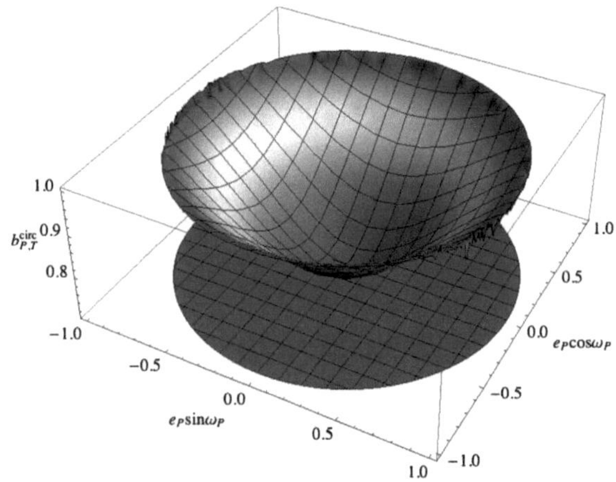

Fig. 4.3 If one uses the circular expressions, the retrieved impact parameter is heavily biased by eccentricity. In this example, the true value of $b_{P,T}$ is $\sqrt{0.5}$ but the introduction of eccentricity causes $b_{P,T}$ to be overestimated

$$[b_{P,T}^{\rm circ}]^2 = 1 + p^2 + 2p$$
$$\times \left(\frac{\sin^2\left[\frac{\varrho_{P,T}^2}{\sqrt{1-e_P^2}} \arcsin\left(\frac{\sqrt{(1-p)^2-b_{P,T}^2}}{(a_P/R_*)\varrho_{P,T}\sin i_P}\right)\right] + \sin^2\left[\frac{\varrho_{P,T}^2}{\sqrt{1-e_P^2}} \arcsin\left(\frac{\sqrt{(1+p)^2-b_{P,T}^2}}{(a_P/R_*)\varrho_{P,T}\sin i_P}\right)\right]}{\sin^2\left[\frac{\varrho_{P,T}^2}{\sqrt{1-e_P^2}} \arcsin\left(\frac{\sqrt{(1-p)^2-b_{P,T}^2}}{(a_P/R_*)(\varrho_{P,T}\sin i_P}\right)\right] - \sin^2\left[\frac{\varrho_{P,T}^2}{\sqrt{1-e_P^2}} \arcsin\left(\frac{\sqrt{(1+p)^2-b_{P,T}^2}}{(a_P/R_*)\varrho_{P,T}\sin i_P}\right)\right]} \right) \tag{4.68}$$

where it is understood that terms on the right-hand side with $b_{P,T}$ in them refer to the true impact parameter, $b_{P,T} = (a_P/R_*)\varrho_{P,T}\cos i_P$. This function is plotted in the case of $(a_P/R_*) = 10$, $b_{P,T}^2 = 0.5$ and $p = 0.1$ in Fig. 4.3. Making small-angle approximations, this yields $[b_{P,T}^{\rm circ}]^2 \simeq b_{P,T}^2$. However, for larger $e_P \sin\omega_P$ and $e_P \cos\omega_P$ values, the overall effect is to overestimate $b_{P,T}$ for eccentric orbits.

In addition to the impact parameter, SMO03 proposed that the parameter (a_P/R_*) may be derived using:

$$[(a_P/R_*)^{\rm circ}]^2 = \frac{(1+p)^2 - [b_{P,T}^{\rm circ}]^2}{\sin^2(T_{14}^{\rm circ}\pi/P_P)} + [b_{P,T}^{\rm circ}]^2 \tag{4.69}$$

If one uses the assumption $b_{P,T}^{\rm circ} \simeq b_{P,T}$, then this equation yields:

$$[(a_P/R_*)^{\rm circ}]^2 = b_{P,T}^2 + [(1+p^2) - b_{P,T}^2]\csc^2\left[\frac{\varrho_{P,T}^2}{\sqrt{1-e_P^2}} \arcsin\left(\frac{\sqrt{(1+p)^2 - b_{P,T}^2}}{(a_P/R_*)\varrho_{P,T}\sin i_P}\right)\right] \tag{4.70}$$

with small-angle approximations, this simplifies to:

$$(a_P/R_*)^{\rm circ} \simeq (a_P/R_*)\sqrt{\varrho_{P,T}^2\cos^2 i_P + \frac{(1-e_P^2)\sin^2 i_P}{\varrho_{P,T}^2}} \tag{4.71}$$

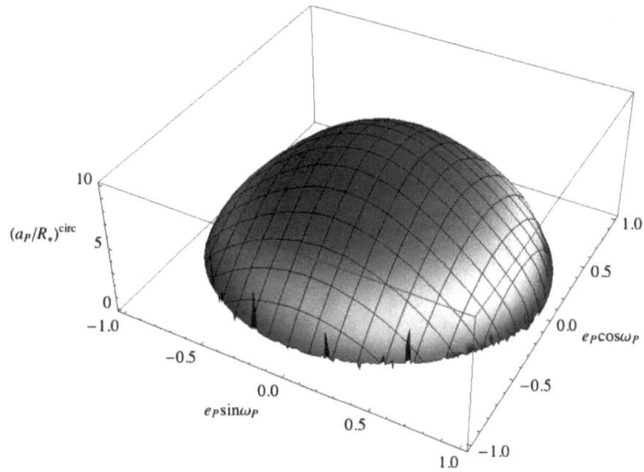

Fig. 4.4 If one uses the circular expressions, the retrieved value of (a_P/R_*) is heavily biased by eccentricity. In this example, the true value of (a_P/R_*) is 10 but the introduction of eccentricity causes (a_P/R_*) to be underestimated

The term inside the square root goes to unity for circular orbits, as expected. The deviation in (a_P/R_*) can be seen to become quite significant for eccentric orbits, as seen in Fig. 4.4 where the exact expression for (4.71) is plotted. This will have significant consequences for the next parameter, the stellar density. As seen earlier, stellar density is related to (a_P/R_*) by manipulation of Kepler's Laws (see (4.41)):

$$\begin{aligned}\rho_* &= \frac{3\pi}{GP_P^2}(a_P/R_*)^3 - p^3\rho_P \\ &\simeq \frac{3\pi}{GP_P^2}(a_P/R_*)^3 \end{aligned} \tag{4.72}$$

where the approximation is made using the assumption $p \ll 1$. It can therefore be seen that:

$$\rho_*^{\text{circ}} \simeq \rho_* \left[\varrho_{P,T}^2 \cos^2 i_P + \frac{(1-e_P^2)\sin^2 i_P}{\varrho_{P,T}^2} \right]^{3/2} \tag{4.73}$$

$$\rho_*^{\text{circ}} \simeq \rho_* \Psi_P = \rho_* \left[\frac{(1+e_P \sin\omega_P)^3}{(1-e_P^2)^{3/2}} \right] \tag{4.74}$$

where in the second line I have assumed that $i_P \simeq \pi/2$. A series expansion of Ψ_P into first order of e_P yields $\Psi_P \simeq 1 + 3e_P \sin\omega_P + \mathcal{O}[e_P^2]$. So observers neglecting an eccentricity of $e_P \sim 0.1$ may alter the stellar density by 30%. As an example, if one decreased the density of a solar type G2V star by 30%, the biased average stellar density would be more consistent with a star of spectral type K0V. Indeed, asteroseismologically determined stellar densities of transiting systems could be used to infer Ψ_P.

This density bias, which is plotted in Fig. 4.5, could be extremely crucial in the search for transiting planets. Many discovery papers of new transiting planets

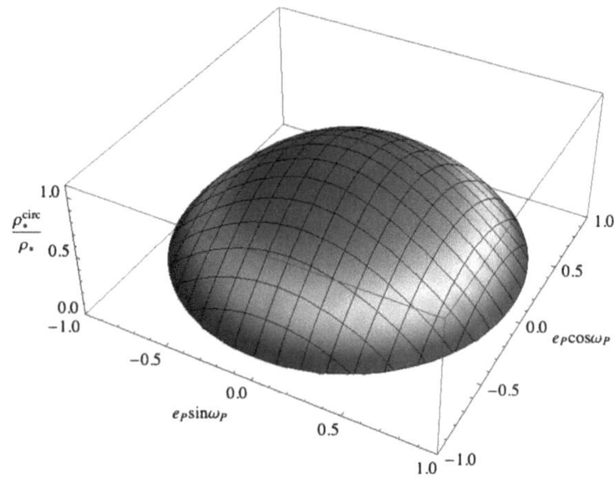

Fig. 4.5 If one uses the circular expressions, the retrieved value of ρ_* is heavily biased by eccentricity. In this example, the true value of ρ_* is one $\rho_\odot$ but the introduction of eccentricity causes ρ_* to be underestimated

have only sparse radial velocity data and usually no secondary eclipse measurement. As a result, the uncertainty on the eccentricity is very large.

Critically, planets are often accepted or rejected as being genuine or not on the basis of this light curve derived stellar density. If the light curve derived stellar density is very different from the combination of stellar evolution and spectroscopic determination, these candidates are generally regarded as unphysical. This method of discriminating between genuine planets and blends, which may mimic such objects, was proposed by Seager and Mallén-Ornelas [4] (see Sect. 6.3 of SMO03) but crucially is only applicable for circular orbits.

Since the typical upper limit on e_P is around 0.1 in discovery papers, then the light curve derived stellar density also has a maximum possible error of $\sim$30%. In practice, the uncertainty on e_P will result in a larger uncertainty in ρ_*. Typical procedure is to fix $e_P = 0$ if the radial velocity data is quite poor, despite the fact the upper limit on $e_P \sim 0.1$. As a result, the posterior distribution of ρ_* would be artificially narrow and erroneous if $e_P \neq 0$. I propose that global fits should allow e_P to vary when analyzing radial velocity and transit photometry, as well as a fixed $e_P = 0$ fit for comparison. This would allow the full range of possible eccentricities to be explored, which would result in a broader and more accurate distribution for $b_{P,T}$, (a_P/R_*) and critically ρ_*.

4.3.6 Application to Multiple Transiting Planet Systems

One interesting application of these ideas comes with the multi-planet transiting systems being found by *Kepler*. Recently, Steffen et al. [14] announced the discovery of five multiple transiting planet candidate systems. The planets have not been confirmed (and thus are "candidate" at the time of writing) because the targets are too faint for RV follow-up. Nevertheless, in such cases, one can assess the ratio of Ψ_P

between the two planets, dubbed with subscripts "b" and "c":

$$\frac{\rho_*^{\mathrm{circ}}(\text{ using planet b})}{\rho_*^{\mathrm{circ}}(\text{using planet c})} = \frac{\Psi_b}{\Psi_c} = \left[\frac{1 + e_b \sin\omega_b}{1 + e_c \sin\omega_c}\right]^3 \left[\frac{1 - e_c^2}{1 - e_b^2}\right]^{3/2} \tag{4.75}$$

The $\sin\omega_P$ functions must be in the range of -1 to $+1$ only, and so it is possible to consider the maximum and minimum ratios which bound the orbits:

$$\left(\frac{\Psi_b}{\Psi_c}\right)_{\min} \leq \left(\frac{\Psi_b}{\Psi_c}\right) \leq \left(\frac{\Psi_b}{\Psi_c}\right)_{\max}$$
$$\left(\frac{1 - e_b}{1 + e_b}\right)\left(\frac{1 - e_c}{1 + e_c}\right) \leq \left(\frac{\Psi_b}{\Psi_c}\right)^{2/3} \leq \left(\frac{1 + e_b}{1 - e_b}\right)\left(\frac{1 + e_c}{1 - e_c}\right) \tag{4.76}$$

Re-writing e_c as $e_b(e_c/e_b)$ and then expanding to first order[7] in e_b on both sides of the inequality, one obtains a symmetric solution on either side which can be consequently re-arranged to:

$$e_b + e_c \geq \frac{1}{2}\left|\left(\frac{\Psi_b}{\Psi_c}\right)^{2/3} - 1\right| \tag{4.77}$$

where it should be understood that the above is only an approximate formula due to the first-order expansion in e_b. To test the accuracy of (4.77), which I label as the Ψ-inequality, I generated some random values for e_b, ω_b, e_c and ω_c. The ω_P values have uniform distributions between 0 and 2π and the e_P values have uniform distributions between 0 and $e_{\max}$. I generated these random values 10^6 times and tested if the inequality in (4.77) was true or not each time. As an example, using $e_{\max} = 0.25$, the inequality is true in 91.9% of all of the Monte Carlo simulations. In Fig. 4.6, I show the percentage of trials for which the inequality is correct as a function of $e_{\max}$, which reveals that the Ψ-inequality provides useful eccentricity constraints in the absence of any other information and is $\geq$90% reliable for $e_{\max} \leq 0.30$.

I also tried using a potentially more realistic non-uniform distribution eccentricity distribution using a mixture of an exponential and a Rayleigh distribution (see [15] and [16]):

$$e_P(x) = \alpha\lambda \exp(-\lambda x) + (1 - \alpha)\frac{x}{\sigma_e^2}\exp(-x^2/2\sigma_e^2) \tag{4.78}$$

The values of the constants were found by fitting the distribution of eccentricities in known multi-planet systems measured from radial velocity surveys using only systems with measured eccentricities, which find $\alpha = 0.38$, $\lambda = 15$ and $\sigma_e = 0.17$ [14]. Finally, this distribution can produce values of e_P greater than unity, and so I ignored any simulations where $e_P > e_{\max}$ for either planet. Using $e_{\max} = 1$, I found that

[7] A first order expansion in eccentricity can be justified by the fact multi-planet systems are rarely permitted to have dynamically stable highly eccentric orbits for either body.

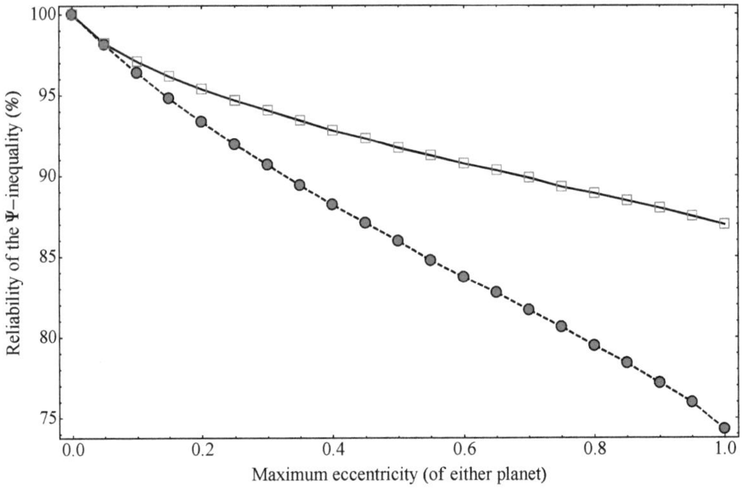

Fig. 4.6 Assuming circular orbits, one may derive two values for the stellar density for a system with two transiting planets. The ratio of these densities constrains the sum of the eccentricities via the Ψ-inequality given in (4.77). The reliability of this inequality is plotted here as a function of the maximum allowed eccentricity, assuming a uniform distribution in e_P (dashed line) and also a non-uniform physically motivated distribution (solid line). The reliability is $\geq 90\%$ for $e_P \leq 0.30$ for the uniform case, and $e_P \leq 0.65$ for the non-uniform case

87.0% of simulations agreed with the inequality presented in (4.77), and $\geq$90% agree for $e_{\mathrm{max}} \leq 0.65$ (Fig. 4.6 shows dependency of this percentage with e_{max}).

Additional constraints may come from dynamical modelling of the orbits to explore the stable regimes of e_P and ω_P for each planet. Put together, these methods raise the chances of characterizing the eccentricity of such systems.

4.3.7 Generalized Parameter Retrieval

In the previous subsection, I discussed how using the circular expressions to derive (a_P/R_*) and ρ_* can lead to severe errors for even mildly eccentric systems. Here, I present expressions which will recover excellent approximate values for $b_{P,T}$, (a_P/R_*) and ρ_*. The new equations (derived using the $\tilde{T}^{\mathrm{one}}$ model and thus given the same superscript notation) are given by:

$$[b_{P,T}^{\mathrm{one}}]^2 = \frac{(1-p)^2 - \dfrac{\sin^2[(T_{23}^{\mathrm{one}}\pi\sqrt{1-e_P^2})/(P_P\varrho_{P,T}^2)]}{\sin^2[(T_{14}^{\mathrm{one}}\pi\sqrt{1-e_P^2})/(P_P\varrho_{P,T}^2)]}(1+p)^2}{1 - \dfrac{\sin^2[(T_{23}^{\mathrm{one}}\pi\sqrt{1-e_P^2})/(P_P\varrho_{P,T}^2)]}{\sin^2[(T_{14}^{\mathrm{one}}\pi\sqrt{1-e_P^2})/(P_P\varrho_{P,T}^2)]}} \tag{4.79}$$

$$[(a_P/R_*)^{\text{one}}]^2 = \frac{(1+p^2) - b_{P,T}^2}{\varrho_{P,T}^2 \sin^2[(T_{14}^{\text{one}}\pi\sqrt{1-e_P^2})/(P_P\varrho_{P,T}^2)]} + \frac{b_{P,T}^2}{\varrho_{P,T}^2} \tag{4.80}$$

$$\rho_*^{\text{one}} = \frac{3\pi}{GP_P^2}[(a_P/R_*)^{\text{one}}]^3 - p^3\rho_P \tag{4.81}$$

These expressions can be shown to reduce down to the original SMO03 derivations if $e_P \to 0$ ((4.7) and (4.8) of SMO03). The new stellar density parameter may be used with floating e_P and ω_P values to correctly estimate the probability distribution of this critical parameter.

It is also possible to write down the retrieval of the parameters in terms of the duration $\tilde{T}$, which was defined in Sect. 4.2.3, and the ingress/egress duration. Due to the approximations made in the derivation of T_{xy}^{one}, the ingress and egress have an equivalent duration, as was seen for circular orbits. Therefore, T_{12} and $\tilde{T}$ form a complete set which can be used instead of the T_{14} and T_{23} set. With this alternate set, the retrieved parameters may be shown to be:

$$\tilde{\mathfrak{T}}^{\text{one}} = \frac{2\pi\sqrt{1-e_P^2}\tilde{T}^{\text{one}}}{P_P\varrho_{P,T}^2} \tag{4.82}$$

$$\mathfrak{T}_{1,2}^{\text{one}} = \frac{2\pi\sqrt{1-e_P^2}T_{12}^{\text{one}}}{P_P\varrho_{P,T}^2} \tag{4.83}$$

$$\begin{aligned}(a_P/R_*)^2 =& \Big[\varrho_{P,T}^2\Big(2p^2\cos\tilde{\mathfrak{T}}^{\text{one}} + (p^2+1)\cos 2\tilde{\mathfrak{T}}^{\text{one}} - \cos 2\mathfrak{T}_{1,2}^{\text{one}} + p^2\Big) \\ & - \sqrt{2}p\varrho_{P,T}^2 \csc^2\mathfrak{T}_{1,2}^{\text{one}} \cos^2\frac{\tilde{\mathfrak{T}}^{\text{one}}}{2} \sin 2\mathfrak{T}_{1,2}^{\text{one}} \\ & \times \sqrt{p^2 - (p^2-4)\cos 2\mathfrak{T}_{1,2}^{\text{one}} - 4\cos 4\mathfrak{T}_{1,2}^{\text{one}}}\Big] \\ & \times \Big[\varrho_{P,T}^4\Big(\cos 2\tilde{\mathfrak{T}}^{\text{one}} - \cos 2\mathfrak{T}_{1,2}^{\text{one}}\Big)\Big]^{-1}\end{aligned} \tag{4.84}$$

$$\begin{aligned}b_{P,T}^2 =& \Big[\varrho_{P,T}^2\big(-2p^2\cos\tilde{\mathfrak{T}}^{\text{one}} + (p^2+1)\cos 2\tilde{\mathfrak{T}}^{\text{one}} + p^2 - 1\big) \\ & + \sqrt{2}p\varrho_{P,T}^2 \tan^2\frac{\tilde{\mathfrak{T}}^{\text{one}}}{2} \csc^2\mathfrak{T}_{1,2}^{\text{one}} \cos^2\frac{\tilde{\mathfrak{T}}^{\text{one}}}{2} \sin 2\mathfrak{T}_{1,2}^{\text{one}} \\ & \times \sqrt{p^2 - (p^2-4)\cos 2\mathfrak{T}_{1,2}^{\text{one}} - 4\cos 2\tilde{\mathfrak{T}}^{\text{one}}} + 2\varrho_{P,T}^2\sin^2\mathfrak{T}_{1,2}^{\text{one}}\Big] \\ & \times \big[\varrho_{P,T}^2\big(\cos 2\tilde{\mathfrak{T}}^{\text{one}} - \cos 2\mathfrak{T}_{1,2}^{\text{one}}\big)\big]^{-1}\end{aligned} \tag{4.85}$$

where the first two lines give a necessary substitution.

4.3.8 Application to Light Curve Fitting

In fitting transit light curves, one is trying to calculate the parameters τ_T, p, (a_P/R_*) and $b_{P,T}$. One can fit for these parameters directly but the strong correlation between (a_P/R_*) and $b_{P,T}$ severely slows down such algorithms [7]. Consequently, it is preferable to use an alternative pair of parameters over (a_P/R_*) and $b_{P,T}$.

To understand what alternative pair should be used, one must analyze the light curve morphology which can be done by approximating the transit light curve as a trapezoid and calculating the various covariances between the different parameters. Carter et al. [7] showed that this exercise reveals that T_{12} and $\tilde{T}$ exhibit much lower correlations than $b_{P,T}$ and (a_P/R_*). One must also consider the effect of whether the fitted parameters have uniform priors or not, which is also a desideratum.

Various sets exist in the literature e.g. Winn et al. [17] advocate $\tilde{T}$ and $b_{P,T}$; Bakos et al. [3] propose $(\zeta/R_*) = 2/\tilde{T}$ and $b_{P,T}^2$. The former set is generally preferable to the latter due to the non-uniform prior which exists for $b_{P,T}^2$ [18], but nevertheless the Bakos et al. [3] set is far better than using (a_P/R_*) and $b_{P,T}$. However, Bakos et al. [3] used the TS05 expression for $\tilde{T}$. By using an improved estimate for the duration, one would expect the correlations to further decrease, since the offset between the approximation and the true value is lower.

To investigate this hypothesis, I generated an artificial transit and refitted the data using the Markov Chain Monte Carlo (MCMC) method, which allows one to see the various inter-parameter correlations. I used three parameter sets to fit the data. Firstly, I use the physical parameter set $\{\tau_T, p, (a/R_*), b_{P,T}\}$ as a reference, or "control", model. In the second I use $\{\tau_T, p^2, (\zeta/R_*), b_{P,T}^2\}$ and in the third I use $\{\tau_T, p^2, (\Upsilon/R_*), b_{P,T}^2\}$, where I have defined $(\Upsilon/R_*) = 2/\tilde{T}^{\mathrm{one}}$.

Since the TS05 formula does not reduce down to the SMO03 equation, the formula will perform worse than T^{one} for even circular orbits (or near-circular), which are the most common type of planet in the exoplanet transiting catalogue (see http://www.exoplanet.eu). I therefore consider a hot-Jupiter on a circular orbit with a planet-star separation of $(a_P/R_*) = 3.5$ from a Sun-like star ($P_P = 0.76$ days). I choose a near-grazing transit with $b_{P,T} = 0.9$, corresponding to an orbital inclination of $i_P = 75.1°$. The light curve is generated using the Mandel and Agol [19] algorithm with no limb-darkening and 0.25 mmag Gaussian noise over a 30 sec cadence. The light curve is then passed onto an MCMC fitting algorithm. I also repeated the whole exercise for an eccentric orbit example as well, where I chose the same system parameters as that of the most eccentric transiting planet known at the time of writing, HD 80606b ([17, 20]).

In the MCMC runs, the jump sizes are set to be equal to $\sim 1-\sigma$ uncertainties from a preliminary short-run. The MCMC is started from 5-σ's away from the solution for each parameter, utilizing 500,000 trials with a 100,000 burn-in time. I then compute the cross-correlations between the various parameters in trials which are within $\Delta\chi^2 = 1$ of $\chi^2|_{\mathrm{best}}$ (errors rescaled such that $\chi^2|_{\mathrm{best}}$ equals number of data points minus the degrees of freedom). The inter-parameter correlations are computed and used to construct correlation matrices for each parameter fitting set. As an example,

Table 4.1 For several proposed light curve fitting parameter sets (left column), I calculate the inter-parameter correlation matrices in the examples of (1) a hypothetical near-grazing hot-Jupiter on a circular orbit (2) a system similar to the eccentric planet HD 80606b.

Parameter Set	$\sum_{i=1}^{4} \lvert M_{i,i} - 1 \rvert$
Circular orbit example	
$\{\tau_T, p, (a_P/R_*), b_{P,T}\}$	2.19333
$\{\tau_T, p^2, (\zeta/R_*), b_{P,T}^2\}$	1.71236
$\{\tau_T, p^2, (\Upsilon/R_*), b_{P,T}^2\}$	1.32974
Eccentric orbit example	
$\{\tau_T, p, (a_P/R_*), b_{P,T}\}$	2.46676
$\{\tau_T, p^2, (\zeta/R_*), b_{P,T}^2\}$	1.56816
$\{\tau_T, p^2, (\Upsilon/R_*), b_{P,T}^2\}$	1.56799

I diagnolize the correlation matrices to give M and then quantify the departure from a perfectly optimal parameter set (right column), where it is understood that 0 corresponds to optimal and larger values correspond to greater inter-parameter correlations

the correlation matrix for the $\{\tau_T, p, (a_P/R_*), b_{P,T}\}$ parameter set is given by:

$$\mathrm{Corr}(\{\tau_T, p, (a_P/R_*), b_{P,T}\}, \{\tau_T, p, (a_P/R_*), b_{P,T}\})$$
$$= \begin{pmatrix} 1 & \mathrm{Corr}(\tau_T, p) & \mathrm{Corr}(\tau_T, (a_P/R_*)) & \mathrm{Corr}(\tau_T, b_{P,T}) \\ \mathrm{Corr}(p, \tau_T) & 1 & \mathrm{Corr}(p, (a_P/R_*)) & \mathrm{Corr}(p, b_{P,T}) \\ \mathrm{Corr}((a_P/R_*), \tau_T) & \mathrm{Corr}((a_P/R_*), p) & 1 & \mathrm{Corr}((a_P/R_*), b_{P,T}) \\ \mathrm{Corr}(b_{P,T}, \tau_T) & \mathrm{Corr}(b_{P,T}, p) & \mathrm{Corr}(p, (a_P/R_*)) & 1 \end{pmatrix}$$

I then calculated the semi-principal axes of correlation ellipsoid by diagnolizing the matrices. For a completely optimal parameter set, this diagnolized matrix would be the identity matrix. I quantify the departure of each proposed parameter set from the identity matrix by calculating $\sum_{i=1}^{4} |M_{i,i} - 1|$ where $\mathbf{M}$ is the diagnolized correlation matrix. The results are shown in upper half of Table 4.1.

The correlations of the physical parameter set are predictably very large, in particular between (a_P/R_*) and $b_{P,T}$, which approaches negative unity. The improvement of T^{one} over T^{TS05} is evident by examining the effect of modifying (ζ/R_*) to (Υ/R_*), which produces a clear improvement in the corresponding correlations, as seen in Fig. 4.7.

4.3.9 Secular Transit Duration Variations

Transit duration variation (TDV) can occur in two possible formats: (1) periodic change and/or (2) secular change (or apparently secular). Periodic changes in the duration of a transit will be discussed later in Chap. 6.

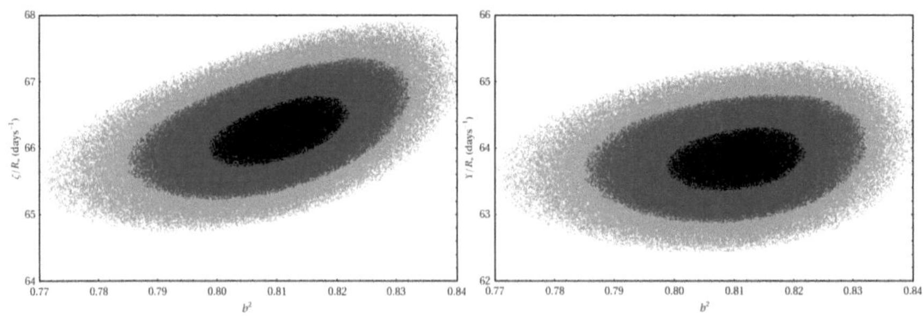

Fig. 4.7 Comparison of the correlations between (ζ/R_*) against $b^2_{P,T}$ (left panel) and (Υ/R_*) against $b^2_{P,T}$ (right panel). Data comes from fitting a synthetic hot-Jupiter light curve on a circular, near-grazing orbit with an MCMC algorithm. The (Υ/R_*) parameter provides two-fold lower correlation and preserves the ability to be inversely mapped to more physical parameters (see (4.81)). The three different types of shading represent the 1-σ, 2-σ and 3-σ confidence regions

Secular changes in transit duration can be caused by numerous possible scenarios. For example, Jordán and Bakos [11] showed that apsidal precession would induce changes in $\tilde{T}$ and used the TS05 equation to predict the size of these changes. As was already demonstrated a better formulation for $\tilde{T}$ is possible; I will here present an improved equation for the rate of change in $\tilde{T}$ due to apsidal precession, or essentially changes in ω_P.

Jordán and Bakos [11] argued that apsidal precession can be caused by stellar oblateness, general relativistic effects and/or a perturbing planet. Additionally, Murray and Dermott [21] showed that, in general, nodal precession should always accompany apsidal precession, leading to changes in the orbital inclination angle, i_P. These changes lead to another form of secular TDV.

Additionally, I consider here that falling planets, such as proposed for WASP-18b [22], would experience a changing semi-major axis, a_P, leading to another form of secular TDV. Finally, I will consider the effect of varying the orbital eccentricity. All four possible TDVs will be derived here using $\tilde{T}^{\text{one}}$, since T^{one} has demonstrated the greatest accuracy in numerical tests (see Sect. 4.3.3) and $\tilde{T}$, in general, is the most useful definition of the duration for TDV studies [7].

4.3.9.1 Apsidal Precession

Apsidal precession is the precession of the argument of periapse over time and it may be induced from several different effects including:

- General relativistic effects ([23, 24])
- Rotational quadrupole bulges on the planet [25]
- Tides raised on the planet and the star [25]

- Stellar quadrupole moment [21]
- Kozai mechanism [26]
- Perturbing planets ([21, 27, 28])

Out of these examples, planets on nearby orbits of masses $\geq M_\oplus$ are expected to produce the largest effect. Thus the detection of apsidal precession could actually be used to infer the presence of Earth-mass planets.

As Jordán and Bakos [11] noted, apsidal precession should cause a change in the transit duration and in order to estimate the magnitude of this effect, the authors differentiated T^{TS05} with respect to ω_P. Having shown the T^{one} offers substantial improvement over the T^{TS05} in the previous section, I am here able to provide an improved estimate for the secular TDV caused by apsidal precession:

$$\frac{\partial \tilde{T}^{\text{one}}}{\partial \omega_P} = \frac{P_P}{\pi} \frac{e_P \varrho_{P,T}^3 \cos \omega_P}{(1-e_P^2)^{3/2}} \left(\frac{1}{\sqrt{1-b_{P,T}^2}\sqrt{(a_P/R_*)^2 \varrho_{P,T}^2 - 1}} - 2 \arcsin\left(\frac{\sqrt{1-b_{P,T}^2}}{(a_P/R_*)\varrho_{P,T} \sin i_P} \right) \right) \tag{4.86}$$

One can see that there are two terms counter-acting in the derived quantity. The two terms can be understood to originate from the planet-star separation changing as a result of the precession, which has two effects: (1) decreasing the planet-star separation causes a near-grazing transit's impact parameter to decrease and thus increases $\tilde{T}$ (the first term) (2) decreasing the planet-star separation causes the tangential orbital velocity to increase and thus decreases $\tilde{T}$ (the second term). The $\cos \omega_P$ term outside of the brackets determines the sign of which term causes an increase and which to decrease.

Kopal [1]showed that the two effects approximately cancel out for $b_{P,T} \simeq 1/\sqrt{2}$. The Kopal [1] derivation is quite different for the ones produced in this paper. It is done by first solving for the mid-eclipse time by a series expansion of the differential of the planet-star separation with respect to true anomaly, disregarding terms in $\sin^3 i_P$ or higher. The duration between the primary and secondary occultation is then solved for in another series expansion in first order of e_P. Nevertheless, setting $b_{P,T}$ to this value, the terms inside the bracket of (4.86) become:

$$\frac{\sqrt{2}}{\sqrt{(a_P/R_*)^2 \varrho_{P,T}^2 - 1}} - 2 \arcsin\left(\frac{1}{\sqrt{2}\sqrt{(a_P/R_*)^2 \varrho_{P,T}^2 - \frac{1}{2}}} \right) \tag{4.87}$$

Under the condition $a_P \gg R_*$, I find that (4.86) gives $\partial \tilde{T}^{\text{one}}/\partial \omega_P = 0$, in agreement with Kopal [1]. For very close-in orbits, this does not appear to hold.

One may compare this estimate of the apsidal precession to (4.15) of Jordán and Bakos [11], which was found by differentiating the expression of TS05 with respect to ω_P. The difference between the two expression is typically less than 1% across a broad range of parameters. However, if $b_{P,T} \simeq 1/\sqrt{2}$, the difference between the two diverges and can reach 10–100%. Given the sensitivity of both equations to this critical value of $b_{P,T}$, I recommend numerical calculations over analytic approximations if $b_{P,T}$ is known to be close to 0.707.

4.3.9.2 Nodal Precession

Nodal precession causes changes in the orbital inclination of the planetary orbit, which would be a source of secular TDV. The secular theory of Murray and Dermott [21] predicts the rate of inclination change due to a perturbing planet as the nodes precess:

$$\frac{\partial i_P}{\partial t} = -\frac{\partial \omega_P}{\partial t} \Delta\Omega \tag{4.88}$$

where $\Delta\Omega$ is the ascending node of the perturbing planet relative to the ascending node of the transiting planet, measured clockwise on the plane of the sky. Therefore, any occurrence of apsidal precession from a perturbing planet will, in general, be coupled with nodal precession. One may derive the rate of secular TDV from inclination change as before:

$$\frac{\partial \tilde{T}^{\mathrm{one}}}{\partial i_P} = \frac{P_P}{\pi} \frac{\varrho_{P,T}^2 \sqrt{(a_P/R_*)^2 \varrho_{P,T}^2 - 1}}{\tan i_P \sqrt{1-e_P^2}\sqrt{1-b_{P,T}^2}} \tag{4.89}$$

This expression has only one term and therefore one can see that decreasing the inclination towards a more grazing transit always yields a shorter transit duration, and vice versa.

4.3.9.3 Falling Exoplanets

Planetary bodies experience infall towards the host star through tidal dissipation and to a much lesser degree gravitational radiation. The effects increase as the orbit becomes smaller leading to runaway fall-in. Therefore, for very close-in exoplanets, the change in semi-major axis may be detectable. The transit duration will vary as:

$$\frac{\partial \tilde{T}^{\mathrm{one}}}{\partial a_P} = \frac{P_P}{\pi} \frac{\varrho_{P,T}^2}{a_P \sqrt{1-e_P^2}} \left(\frac{3}{2} \arcsin\left(\frac{\sqrt{1-b_{P,T}^2}}{(a_P/R_*)\varrho_{P,T} \sin i_P} \right) - \frac{1}{\sqrt{1-b_{P,T}^2}\sqrt{(a_P/R_*)^2 \varrho_{P,T}^2 - 1}} \right) \tag{4.90}$$

As for apsidal precession, there are two countering components, which are the same as before except for a slightly different constant in front of the first term. This different constant means that the impact parameter at which both effects cancel has now changed to $b_{P,T} \simeq 1/\sqrt{3} = 0.577$. This result could not be found in the previous literature and is of particular interest given the recent discovery of exoplanets on periods of around a day or less, for example WASP-18b [22] with period of 0.94 days and $b_{P,T} = 0.25$.

4.3.9.4 Eccentricity Variation

Irregular satellites are known to exchange orbital inclination and eccentricity through the Kozai mechanism, which roughly conserves the value $\cos I\sqrt{1-e^2}$, where I is the angle to the ecliptic. Variations in e_P may lead to a form of secular TDV as well:

$$\frac{\partial \tilde{T}^{\text{one}}}{\partial e_P} = \frac{P_P}{\pi}\frac{\varrho_{P,T}^3}{(1-e_P^2)^{5/2}}\Bigg[\frac{2e_P + (1+e_P^2)\sin\omega_P}{\sqrt{1-b_{P,T}^2}\sqrt{(a_P/R_*)^2\varrho_{P,T}^2 - 1}} - [3e_P + (2+e_P^2)\sin\omega_P]\arcsin\left(\frac{\sqrt{1-b_{P,T}^2}}{(a_P/R_*)\varrho_{P,T}\sin i_P}\right)\Bigg] \tag{4.91}$$

The two terms here seem to exhibit a more complicated inter-dependency, which is physically based on the same idea of varying the planet-star separation. The balance-point between the two effects occurs for:

$$b_{P,T} \simeq \sqrt{\frac{e_P + \sin\omega_P}{(3e_P + (2+e_P^2)\sin\omega_P}} = \frac{1}{\sqrt{2}} - \frac{e_P}{4\sqrt{2}\sin\omega_P} + \mathcal{O}[e_P]^2 \tag{4.92}$$

Thus to zeroth-order in e_P, the balance point is the same as for apsidal precession.

4.4 Other Important Timings

4.4.1 Transit to Occultation Time

Before moving on, I will discuss other important timings related to transits, aside from the transit minima and the durations. The first one will I consider is the time it takes for a planet to move from the transit minimum to the occultation minimum i.e. $\tau_O - \tau_P$. As was discussed in Sect. 3.3.4, this time difference is given by $0.5P_P + \Delta t_{\text{light}}$ for circular orbits, where Δt_{light} is the light travel time across the system. For an eccentric orbit, things are more complicated and the exact time is given by:

$$(\tau_O - \tau_T) = \frac{P_P}{2\pi}\Delta\mathfrak{M}_{T\to O} + \Delta t_{\text{light}} \tag{4.93}$$

$$\Delta\mathfrak{M}_{T\to O} = \int\limits_{f_P=f_{P,T}}^{f_{P,O}} \frac{(1-e_P^2)^{3/2}}{(1+e_P\cos f_P)^2}\mathrm{d}f_P \tag{4.94}$$

Dealing with the light travel time first, if one assumes a nearly coplanar orbit, then the transit minima occur at the times of the conjunctions, so:

$$\begin{aligned}\Delta t_{\text{light}} &= \frac{Z_P(f_{P,O}) - Z_P(f_{P,T})}{c} \simeq \frac{a_P(\varrho_{P,T}+\varrho_{O,T})}{c}\\ \lim_{i_P\to\pi/2}\Delta t_{\text{light}} &= \frac{2a_P(1-e_P^2)}{c(1-e_P^2\sin^2\omega_P)}\end{aligned} \tag{4.95}$$

For the dynamical time difference, one must deal with the integral once again. In practice, my light curve fitting routine computes the transit and occultation minima using the η-series and then uses (4.44) to compute $\Delta\mathfrak{M}_{T\to O}$. If one must insist on using an approximation though, then $f_{P,T} \simeq (\pi/2 - \omega_P)$ and $f_{P,O} \simeq (3\pi/2 - \omega_P)$ may be used to give:

$$\begin{aligned}\lim_{i_P\to\pi/2}(\tau_O - \tau_T) = &\frac{P_P}{2} + \frac{P_P}{\pi}\left(\arctan\left[\frac{e_P\cos\omega_P}{\sqrt{1-e_P^2}}\right] + \frac{e_P\sqrt{1-e_P^2}\cos\omega_P}{1-e_P^2\sin^2\omega_P}\right)\\ &+ \frac{2a_P(1-e_P^2)}{c(1-e_P^2\sin^2\omega_P)}\end{aligned} \tag{4.96}$$

Another insightful approximation comes from expanding to first order in e_P and ignoring the light travel time, which is usually very small:

$$(\tau_O - \tau_T) \simeq \frac{P_P}{2} + \frac{2P_P}{\pi}\left[e_P\cos\omega_P + \mathcal{O}[e_P^2]\right] \tag{4.97}$$

Therefore, the measurement of the time between the transit and the occultation strongly constrains $e_P\cos\omega_P$.

4.4.2 Transit to Occultation Duration Ratio

To estimate of the ratio between the two eclipse durations, I will proceed with the T^{one} approximations. I will label the occultation duration between the x and y contact points as O_{xy} (thus T corresponds to the transit and O for the occultation). It is trivial to show that:

$$O_{xy}^{\text{one}} = \frac{P_P}{\pi} \frac{\varrho_{P,O}^2}{\sqrt{1-e_P^2}} \arcsin\left[\sqrt{\frac{S_{P*}^2 - b_{P,O}^2}{(a_P/R_*)^2\varrho_{P,O}^2 - b_{P,O}^2}}\right] \tag{4.98}$$

Where $b_{P,O}$ is the impact parameter of the planet for the occultation event, given by $b_{P,O} = (a_P/R_*)\varrho_{P,O} \cos i_P$. Taking the ratio of the tilde definitions for the duration gives an elaborate expression. However, I expand to zeroth-order in $\cos^2 i_P$ and then to first order in e_P, simplifying the result with the assumption $a_P \gg R_*$:

$$\frac{\tilde{T}^{\text{one}}}{\tilde{O}^{\text{one}}} \simeq 1 - 2e_P \sin\omega_P + \mathcal{O}[e_P^2] \tag{4.99}$$

Due to the approximations made, the above expression is only useful for providing a sense of the scaling of the ratio. However, they do allow one to see that the dominant term is $e_P \sin\omega_P$. Therefore, the transit to occultation timing gives $e_P \cos\omega_P$ and the duration ratios give $e_P \sin\omega_P$ suggesting that eccentricity could be measured photometrically with high quality data.

4.4.3 Transit to RV Time

As a final piece to the puzzle, I will discuss the time difference between the transit minimum and the time of the stellar reflex velocity null (determined by radial velocity measurements). The RV-null occurs at exactly the same instant as the transit minimum for circular orbits. However, for eccentric orbits a slight offset is introduced. This may be found by first finding the true anomaly at which the RV-null occurs, by solving:

$$\begin{aligned} \text{RV}(f_P) - \gamma &= -K \sin i_P[e_P \cos\omega_P + \cos(f_P + \omega_P)] = 0 \\ \Rightarrow f_{P,\text{null}} &= -\omega_P + \arccos[-e_P \cos\omega_P] \end{aligned} \tag{4.100}$$

Thus for $e_P = 0$ one has $f_{P,\text{null}} = \pi/2 - \omega_P$, as expected. The disturbance of the RV-null away from the point of inferior conjunction is much greater than that of the transit minimum. Therefore, a good approximation is given by assuming $f_{P,T} \simeq \pi/2 - \omega_P$ for the purposes of computing the time difference between the RV-null and the transit minimum. I will also use the approximation $\Delta\mathfrak{M}_P \sim \Delta f_P$ for simplicity:

$$\begin{aligned} (\tau_{\text{null}} - \tau_T) &\simeq \frac{P_P}{2\pi}[\arccos[-e_P \cos\omega_P] - \pi/2] \\ &= \frac{P}{2\pi}\left[e_P \cos\omega_P + \mathcal{O}[e_P^2]\right] \end{aligned} \tag{4.101}$$

The first-order expansion in e_P, which is presented on the second-line, illustrates how the RV-null time has a strong dependence on $e_P \cos\omega_P$. Indeed, at the first-order level, $(\tau_{\text{null}} - \tau_T) \simeq (\tau_O - \tau_T - 0.5P_P)/4$. Put together, these additional timings permit the joint modelling of transit, occultation and RV data to strongly constrain e_P and ω_P. With these quantities known, reliable estimates for the other parameters, such as i_P and ρ_*, are enabled.

References

1. Z. Kopal, Close binary systems. (1959)
2. J.B. Irwin, The determination of a light-time orbit. Astrophys. J. **116**, 211 (1952). 10.1086/145604
3. G.Á. Bakos, G. Kovács, G. Torres, D.A. Fischer, D.W. Latham, R.W. Noyes, D.D. Sasselov, T. Mazeh, A. Shporer, R.P. Butler, R.P. Stefanik, J.M. Fernández, A. Sozzetti, A. Pál, J. Johnson, G.W. Marcy, J.N. Winn, B. Siphocz, J. Lázár, I. Papp, P. Sári, HD 147506b: a supermassive planet in an eccentric orbit transiting a bright star. Astrophys. J. **670**, 826–832 (2007)
4. S. Seager, G. Mallén-Ornelas, A unique solution of planet and star parameters from an extrasolar planet transit light curve. ApJ **585**, 1038–1055 (2003)
5. I. Baraffe, G. Chabrier, F. Allard, P.H. Hauschildt, Evolutionary models for solar metallicity low-mass stars: mass-magnitude relationships and color-magnitude diagrams. Astron. Astrophys. **337**, 403–412 (1998)
6. S. Yi, P. Demarque, Y.-C. Kim, Y.-W. Lee, C.H. Ree, T. Lejeune, S. Barnes, Toward better age estimates for stellar populations: The Y2 isochrones for solar mixture. Astrophys. J. Suppl. **136**, 417–437 (2001)
7. J.A. Carter, J.C. Yee, J. Eastman, B.S. Gaudi, J.N. Winn, Analytic approximations for transit light-curve observables, uncertainties, and covariances. Astrophys. J. **689**, 499–512 (2008)
8. D.M. Kipping, Transiting planets—light-curve analysis for eccentric orbits. MNRAS **389**, 1383–1390 (2008)
9. B. Tingley, P.D. Sackett, A photometric diagnostic to aid in the identification of transiting extrasolar planets. Astrophys. J. **627**, 1011–1018 (2005)
10. E.B. Ford, S.N. Quinn, D. Veras, Characterizing the orbital eccentricities of transiting extrasolar planets with photometric observations. Astrophys. J. **678**, 1407–1418 (2008)
11. A. Jordán, G.Á. Bakos, Observability of the general relativistic precession of periastra in exoplanets. Astrophys. J. **685**, 543–552 (2008)
12. J.N. Winn, Transits and occultations. ArXiv e-prints, January 2010
13. D.M. Kipping, S.J. Fossey, G. Campanella, On the detectability of habitable exomoons with Kepler-class photometry. MNRAS **400**, 398–405 (2009)
14. J.H. Steffen, N.M. Batalha, W.J. Borucki, L.A. Buchhave, D.A. Caldwell, W.D. Cochran, M. Endl, D.C. Fabrycky, F. Fressin, E.B. Ford, J.J. Fortney, M.J. Haas, M.J. Holman, H. Isaacson, J.M. Jenkins, D. Koch, D.W. Latham, J.J. Lissauer, A.V. Moorhead, R.C. Morehead, G. Marcy, P.J. MacQueen, S.N. Quinn, D. Ragozzine, J.F. Rowe, D.D. Sasselov, S. Seager, G. Torres, W.F. Welsh, Five Kepler target stars that show multiple transiting exoplanet candidates. ArXiv e-prints, June 2010
15. M. Jurić, S. Tremaine, Dynamical origin of extrasolar planet eccentricity distribution. Astrophys. J. **686**, 603–620 (2008)
16. N.L. Zakamska, M. Pan, E.B. Ford, Observational biases in determining extrasolar planet eccentricities in single-planet systems. MNRAS, pp. 1566, October 2010. doi: 10.1111/j.1365-2966.2010.17570.x
17. J.N. Winn, A.W. Howard, J.A. Johnson, G.W. Marcy, J.Z. Gazak, D. Starkey, E.B. Ford, K.D. Colón, F. Reyes, L. Nortmann, S. Dreizler, S. Odewahn, W.F. Welsh, S. Kadakia, R.J. Vanderbei, E.R. Adams, M. Lockhart, I.J. Crossfield, J.A. Valenti, R. Dantowitz, J.A. Carter,

The transit ingress and the tilted orbit of the extraordinarily eccentric exoplanet HD 80606b. Astrophys. J. **703**, 2091–2100 (2009)
18. D.M. Kipping, G.A. Bakos, J. Hartman, G. Torres, A. Shporer, D.W. Latham, G. Kovacs, R.W. Noyes, A.W. Howard, D.A. Fischer, J.A. Johnson, G.W. Marcy, B. Beky, G. Perumpilly, G.A. Esquerdo, D.D. Sasselov, R.P. Stefanik, J. Lazar, I. Papp, P. Sari, HAT-P-24b: an inflated hot-Jupiter on a 3.36d period transiting a hot, metal-poor star. ArXiv e-prints, August 2010
19. K. Mandel, E. Agol, Analytic light curves for planetary transit searches. Astrophys. J. Lett. **580**, L171–L175 (2002)
20. S.J. Fossey, I.P. Waldmann, D.M. Kipping, Detection of a transit by the planetary companion of HD 80606. MNRAS **396**, L16–L20 (2009)
21. C.D. Murray, S.F. Dermott, Solar System Dynamics. (Cambridge University Press, Cambridge, 1999)
22. C. Hellier, D.R. Anderson, A.C. Cameron, M. Gillon, L. Hebb, P.F.L. Maxted, D. Queloz, B. Smalley, A.H.M.J. Triaud, R.G. West, D.M. Wilson, S.J. Bentley, B. Enoch, K. Horne, J. Irwin, T.A. Lister, M. Mayor, N. Parley, F. Pepe, D.L. Pollacco, D. Segransan, S. Udry, P.J. Wheatley, An orbital period of 0.94days for the hot-Jupiter planet WASP-18b. Nature **460**, 1098–1100 (2009)
23. A. Einstein, A. Die Feldgleichungen der Gravitation. Sitzungsberichte der Königlich Preußischen Akademie der Wissenschaften (Berlin), 844–847 (1915)
24. A. Pál, B. Kocsis, Periastron precession measurements in transiting extrasolar planetary systems at the level of general relativity. MNRAS **389**, 191–198 (2008)
25. T.E. Sterne, Apsidal motion in binary stars. MNRAS **99**, 451–462 (1939)
26. Y. Kozai, Secular perturbations of asteroids with high inclination and eccentricity. Astrophys. J. **67**, 591 (1962)
27. J. Miralda-Escudé, Orbital perturbations of transiting planets: a possible method to measure stellar quadrupoles and to detect earth-mass planets. Astrophys. J. **564**, 1019–1023 (2002)
28. J.S. Heyl, B.J. Gladman, Using long-term transit timing to detect terrestrial planets. MNRAS **377**, 1511–1519 (2007)

Chapter 5
Transit Distortions

Nothing is as simple as we hope it will be
Jim Horning

5.1 Introduction

The properties and morphology of the transit light curve have now been established, provided assumptions A0–A10 are in effect. In Sect. 3.3.7, I discussed the consequences of breaking each of the assumptions and how frequent and significant the invalidation of each assumption was. In that section, it was concluded that three assumptions, in particular, were likely to be frequently broken and lead to significant distortion to the transit light curve. These are A0)—a uniform source star; A4/A5)—the planet emits no flux/there are no background luminous objects; A10)—the integration times are small. A0 can be compensated for by using the Mandel and Agol [1] code to account for limb darkening and generally avoid observing spotty stars. In this section, I will discuss the consequences of breaking A4/A5 (Sect. 5.2) and A10 (Sect. 5.3) in more detail and present methods for compensating for each distortion. These two sections are based upon the papers Kipping and Tinetti [2] and Kipping [3] respectively.

5.2 Blending and Nightside Pollution

5.2.1 Introduction

During the early days of exoplanetary science, the dedicated transit space missions now available did not exist and the pioneers of those early times sought the first transit using off-the-shelf instrumentation and small-aperture telescopes. It was therefore not surprising that the first transit light curve was obtained in the visible wavelength

D. M. Kipping, *The Transits of Extrasolar Planets with Moons*, Springer Theses,
DOI: 10.1007/978-3-642-22269-6_5,

range [4, 5] and for the subsequent few years this was established as the normal practice in later observations and surveys (e.g. Brown et al. [6], Bakos et al. [7], Pollacco et al. [8]), mostly down to the cost efficiency of the available CCDs (see Sect. 1.3.6 for a more detailed account of the first discovery).

As transit measurements became routine and staggering photometric quality became available with space-based observatories like HST [6], it became clear that improved models of the transit light curve were required. In response to the growing need for accurate parameterization of light curves, several authors produced equations modelling the light curve behaviour, such as Mandel and Agol [1] and Seager and Mallén-Ornelas [9], both of which have been discussed extensively earlier in this thesis (see Sects. 3.3.6 and 4.2 respectively).

In the last few years, the value of infrared measurements of transiting systems has become apparent with numerous pioneering detections; emission from a transiting planet [10], emission from a non-transiting planet [11], an exoplanetary spectrum [12], detection of water in the atmosphere of a hot-Jupiter [13], methane [14] and more recently carbon dioxide [15]. More details on the use of transmission spectroscopy as a tool for detecting molecular species can be found in Seager and Sasselov [16] and Tinetti and Beaulieu [17]. With JWST set to replace HST in the next decade, one can expect an abundance of high-precision infrared transits to be observed in order to detect more molecular species, perhaps including biosignatures [18]. In this section, I discuss the consequences of significant nightside planetary emission on precise infrared transit light curves. Nightside emission invalidates one of the original assumptions made both earlier in this thesis (assumption A4, Sect. 3.3.1) and in the models of Seager and Mallén-Ornelas [9] and Mandel and Agol [1]: "The planet emits no flux". In the case of hot-Jupiter systems, the nightside of the planet is hot and thus flux-emitting. This additional flux can be considered as a blend, but from the planet itself, i.e. a *self-blend*.

Conceptually, it is very easy to see that this will cause mid-infrared transit depth measurements (and to a lesser degree in the visible and near-infrared range) to become underestimates of the true depth. The reason is that there are two sources of flux, the star and the planet, and only one of these is being occulted, whilst the other is the blend source. This is highly analogous to the case of a nearby companion star which is not spatially resolved in the point-spread-function (PSF) and thus induces blending. In this section, I will derive expressions estimating the amplitude of the effect, propose methods for correcting the light curves and apply them to two cases where the nightside emission of an exoplanet has been determined in the mid-infrared.

Although the fundamental effects of both blending and self-blending (or nightside pollution) are the same (i.e. a dilution of the depth) I will here focus more on the nightside pollution effect. This is because this was not previously known until pointed out in Kipping and Tinetti [2] (the paper upon which this section is based) whereas blending is a well-known occurrence and generally corrected for by resolving the companion's light using high resolution imaging [19].[1]

[1] Obviously this is not possible for self-blending.

5.2.2 Derivation

5.2.2.1 Depth Dilution

Let us define the total out-of-transit flux surrounding a transit event, as shown in Fig. 5.1a, to be given by:

$$F_{\mathrm{out,tra}} = F_* + F_P \tag{5.1}$$

where F_* is the total flux received from the star and F_P is the total flux received from the nightside of the planet, over a time interval of dt. Let us assume that the star is a uniform emitter A0 and that both the stellar and planetary total flux are invariable over the timescale of the transit event A5. One may then write down the flux during a transit as a function of the ratio of the radii, p:

$$F_{\mathrm{in,tra}} = (1 - p^2)F_* + F_P \tag{5.2}$$

Note how the flux of the star has been attenuated as a result of the eclipse but the planetary flux is still present. The observed transit depth in the flux domain, δ', is defined by:

$$\delta' = \frac{F_{\mathrm{out,tra}} - F_{\mathrm{in,tra}}}{F_{\mathrm{out,tra}}} \tag{5.3}$$

$$= \left(\frac{F_*}{F_* + F_P}\right)\delta \tag{5.4}$$

where δ is the undiluted transit depth. In the case of $F_{\mathrm{P}} \to 0$, the standard equation for the depth is recovered, i.e. $\delta' = \delta = p^2$. For cases where the nightside flux of the planet is non-negligible, the transit depth will therefore be affected. One may re-write (5.4) as $\delta = B_P\delta'$ where I define:

$$\begin{aligned} B_x &= \frac{F_* + F_x}{F_*} \\ \Rightarrow B_P &= \frac{F_* + F_P}{F_*} \end{aligned} \tag{5.5}$$

where the above equation emphasizes how the equations presented here are generally applicable to any kind of blend from a source x. For the nightside emission case, the blend source is generally at a much cooler temperature than the host star and thus the contrast between the two bodies is greater at infrared wavelengths. Consequently, the self-blending has a significant effect on infrared measurements (e.g. see Sect. 5.2.5) but a much lower impact on visible wavelengths (e.g. see Sect. 5.2.4). This makes the inclusion of such an effect paramount since visible and infrared measurements must be considered incommensurable unless this systematic is corrected for.

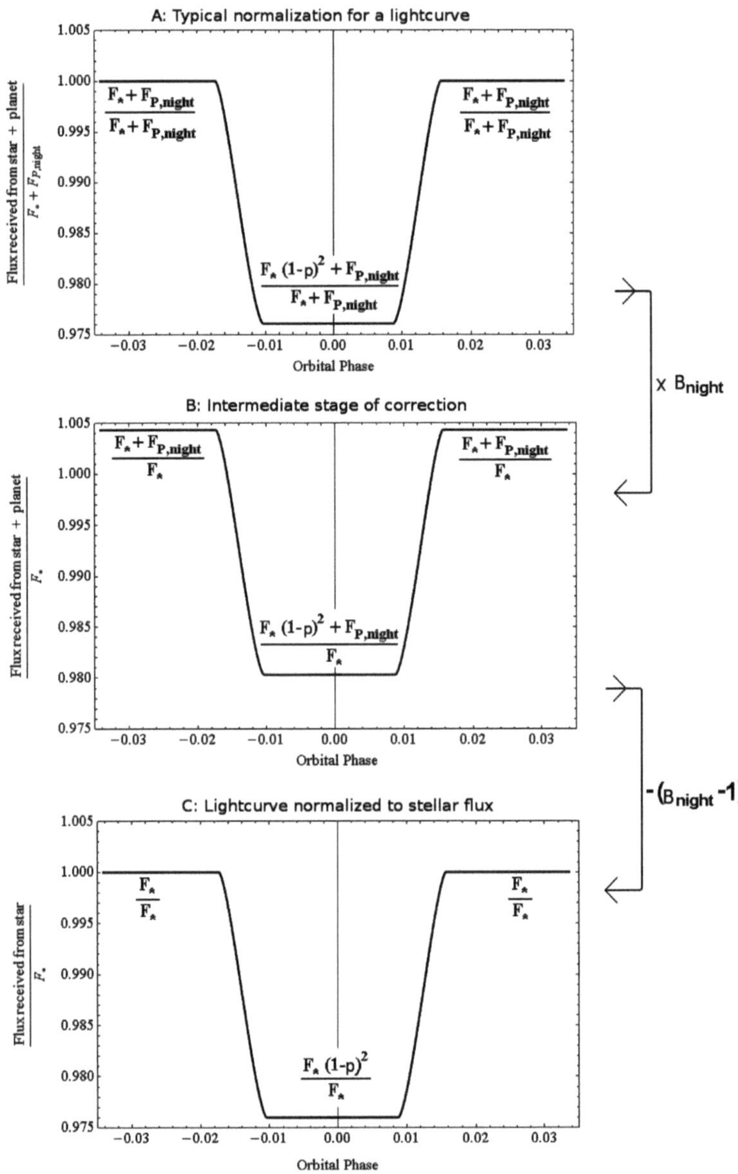

Fig. 5.1 Illustration of the three-stages involved in the corrective procedure, to compensate for the effects of nightside pollution. **a** Typical normalization for a lightcurve, **b** intermediate stage of correction, **c** lightcurve normalized to stellar flux

5.2.2.2 The Consequences for Other Parameters

The largest effect of a blend is to underestimate the transit depth. However, I evaluate here the effect of the nightside blend, or indeed any kind of blend, on the other light curve parameters. For simplicity, I consider here a circular orbit and so I may use the expressions of Seager and Mallén-Ornelas [9].

Earlier in Sect. 4.3.5, I showed how using the circular expressions when the orbit is eccentric leads to the erroneous retrieval of the impact parameter, the scaled orbital semi-major axis and the light curve derived stellar density (see (4.68), (4.71), (4.74)). In a similar manner, I will here show how erroneous the retrieval of these parameters would be if one did not know that a blend was present:

$$\begin{aligned}\left[b_{P,T}^{\text{undiluted}}\right]^2 &= \frac{p^2 - (1 + p^2 - b_{P,T}^2)\sqrt{B_x} + B_x}{B_x} \\ &= b_{P,T}^2 + \frac{1}{2}(1 - b_{P,T}^2 - p^2)(B_x - 1) + \mathcal{O}[(B_x - 1)^2]\end{aligned} \tag{5.6}$$

$$\begin{aligned}[(a_P/R_*)^{\text{undiluted}}]^2 = (a_P/R_*)^2 - \frac{(1 - b_{P,T}^2 - p^2)((a_P/R_*)^2 - (1+p)^2)}{2((1+p)^2 - b_{P,T}^2)}(B_x - 1) \\ + \mathcal{O}[(B_x - 1)^2]\end{aligned} \tag{5.7}$$

Equations (5.6) and (5.7) imply that negating the blending factor causes one to overestimate the impact parameter and underestimate (a_P/R_*). It is important to recall that the light curve derived stellar density is found by taking the cube of (a_P/R_*) and therefore will exacerbate any errors at this stage. Note that both equations give the expected results for $B_x = 1$, i.e. no blend source present.

These expressions have been derived assuming no limb darkening is present, which is typically a very good approximation for the wavelength range we are interested in. However, in reality the incorporation of limb darkening is easily implemented and demonstrated later in Sect. 5.2.5. Therefore, (5.6) and (5.7) should not be used to attempt to correct parameters derived from fits not accounting for nightside pollution, rather they offer an approximate quantification of the direction and magnitude of the expected errors.[2]

5.2.3 *Compensating for the Effect*

5.2.3.1 Empirical Method

In the previous section, it was seen how the undiluted transit depth, δ, and the observed transit depth, δ', are related by the factor B_x. If the blend source was due to a nearby

[2] For this reason, I considered it not worthwhile to derive the full versions of (5.6) and (5.7) which account for eccentricity.

star which was hidden within the PSF (point spread function) during the observations, then it usually possible to evaluate B_x through higher resolution imaging e.g. using adaptic optics [19]. However, if the blend source is from the planet itself, then no feasible improvements in spatial resolution will ever aid in determining B_P.

Fortunately, B_P can be obtained in other ways, specifically through measuring the phase curve of an extrasolar planet (for the first measured example, see Knutson et al. [20]). With such a measurement, the difference between the day and nightside fluxes may be determined and thus (F_P/F_*) can be calculated.

Another possible method is to measure the occultation and transit without the intermediate phase curve information, which would require instruments with extremely stable calibration. For a very inactive star, a highly calibrated instrument could, in principle, measure the nightside flux by measuring just the transits and occultations. This would equate to an absolute calibration accurate to a fraction of the difference between the day and nightside fluxes, estimated to be $\sim 10^{-3}$ for HD 189733b, over one half of the orbital period. I therefore estimate calibration requirements to be at least $\sim 10^{-4}$ during the full ~ 30 h period. Staring telescopes like *Kepler* and *CoRoT* have been shown to reach these precisions [21], but it is the infrared telescopes of *Spitzer* and JWST that are most heavily affected by nightside pollution and these observatories frequently slew around looking at different patches of the sky. After the slewing one requires the target to be at the same centroid position to within a fraction of a pixel. Regardless as to whether the whole phase curve or simply the eclipse-only observations are made, the same method may be used to correct both for the effects of nightside pollution.

A phase curve time series is typically normalized to F_* (see Fig. 3.4 for illustration of the phase curve), which is in contrast to a normal transit measurement, which is normalized to $(F_* + F_P)$ i.e. the local out-of-transit baseline (see Fig. 5.1a). Therefore, for a phase curve, the stellar normalized flux immediately before and after the transit event is equal to B_P, as shown in Fig. 5.1b. It is possible that shifted hot spots on the planetary surface could cause an inequality between the pre and post transit baselines, but in practice the net effect of nightside pollution is very well accounted for by averaging over this time range.

In order to correct a transit light curve, one needs to modify the normalization. In Fig. 5.1, I show the two-step transformation which can achieve this. I consider initially normalizing a light curve using the local baseline as usual for such measurements, as shown in Fig. 5.1a. After this, the two-step correction may be performed, provided the observer has knowledge of B_x. The whole process may be summarized by the following (also illustrated in Fig. 5.1):

1. Normalize fluxes to local out-of-transit baseline, as usual.
2. Multiply all flux values by B_x.
3. Subtract $(B_x - 1)$ from all data points.

In practice, these steps are incorporated into the light curve fitting algorithm directly. In the case of using Monte Carlo based routines for error estimation, B_x may be floated around its best-fit value and corresponding uncertainty. An example of this method is shown in Sect. 5.2.5 for the planet HD 189733b. Defining $I_{j,\mathrm{uncorr}}$

Table 5.1 Best-fit transit parameters for the HD 189733b 8.0 μm primary transit light curve

Method	Depth, p^2 (%)	$\tilde{T}$ (s)	$b_{P,T}$	(a_P/R_*)	i_P (°)	χ^2
(1) Local baseline	2.3824	5127.45	0.66264	8.9121	85.7360	22412.4691
(2) Nightside correction	2.3884	5127.55	0.66204	8.9183	85.7428	22412.4605
Uncertainty	0.0061	8.1	0.0061	0.050	0.054	–
(2)–(1)	+0.0060	+0.10	−0.00060	+0.0062	+0.0068	−0.0086

Data obtained by Knutson et al. [20]. Fits performed for the case of (1) typical normalization of the local baseline (2) correction for the effects of nightside pollution. The number of data points is 22,372

Table 5.2 Best-fit transit parameters of the HD 189733b 24.0 μm primary transit light curve

Method	Depth, p^2 (%)	$\tilde{T}$ (s)	$b_{P,T}$	(a_P/R_*)	i_P (°)	χ^2
(1) Local baseline	2.3980	5072.527	0.61425	9.492	86.290	1260.9125
(2) Nightside correction	2.4085	5072.505	0.6131	9.502	86.300	1260.9092
Uncertainty	0.019	19.7	0.003	0.033	0.022	–
(2)–(1)	+0.011	−0.022	−0.0012	+0.010	+0.010	−0.0033

Data obtained by Knutson et al. [35]. Fits performed for the case of (1) typical normalization of the local baseline (2) correction for the effects of nightside pollution. Number of data points is 1,198

as the locally normalized flux measurement of the jth data point, one may explicitly write down the corrected data point using (5.8).

$$I_{j,\mathrm{corr}} = B_x I_{j,\mathrm{uncorr}} - (B_x - 1) \tag{5.8}$$

I also briefly mention here that the transformation operations on the light curve time series will not only change the transit depth but also provides a more physical transit signal and thus one should expect a slightly lower χ^2 in the final fitting, as indeed is seen later in Tables 5.1 and 5.2.

One caveat with the described method is the possible presence of ellipsoidal variations of the star, which would mix the phase curve signature. For example, Welsh et al. [22] detected ellipsoidal variations in HAT-P-7. Such signals peak at orbital phases of 0.25 and 0.75, whereas a phase curve should peak close to orbital phase 0.5, but can be offset by a small factor due to hot spots. Welsh et al. [22] provide a detailed discussion of modelling both signals and although ellipsoidal variations complicate the analysis, they certainly do not undermine it.

5.2.3.2 Semi-Empirical Method

To correct for the effect of nightside pollution, in an accurate way, I have proposed using phase curve information to obtain B_P, which requires many hours of telescope time. Further, the phase curve should be obtained at every wavelength simultaneously and for every epoch[3] one wishes to measure the transit event at, in order to be sure of a completely reliable correction. I label this resource-intensive method of correcting for

[3] Each epoch should be done in case temporal variability exists in the system.

the effect as the "empirical method". However, it is appreciated that obtaining phase curves at every wavelength for each epoch is somewhat unrealistic and I propose a "semi-empirical method" of achieving the same goal with far fewer resources.

I propose that observations of the phase curve are made at several wavelengths in the infrared; for example the IRAC and MIPS wavelengths of *Spitzer* are very suitable but most of these channels are unfortunately no longer available. The resulting measurements of the nightside flux may be used to compile a spectrum for the purposes of interpolating/extrapolating the emission to other wavelengths which are missing phase curve observations (an example of this will be provided in Sect. 5.2.7). Should higher resolution spectra of the nightside emission become available in the future, one may construct a more sophisticated model as appropriate. Transit observations at different wavelengths may then interpolate/extrapolate the model template to estimate the magnitude of the effect and apply the required correction at any wavelength. This allows one to estimate the blending factor at all wavelengths and times.

Long-term monitoring of the planet may also be necessary in order to ascertain the presence or absence of temporal variability in the system. Ideally, this may be achieved by obtaining phase curves of the exoplanet to measure the nightside flux at regular times. More practically, it could be done by measuring only the occultation, which is the dayside of the exoplanet, at regular times (for example Agol et al. [23]). Any large changes in the nightside flux are likely to be correlated to large changes in the dayside flux too, assuming a constant energy budget for the planet. This second approach would reduce the demands on telescope time by an order of magnitude or more.

5.2.3.3 Non-Empirical Method

The final method I propose here is the least accurate, but requires the fewest resources to implement. If a planet has recently been discovered, no phase curves or even occultations may have been obtained yet. Accordingly, the only avenue available is to estimate the temperature of the nightside through either a simple analytic estimation or a dynamical model of the atmosphere, although the latter may be excessive given the absence of any observational constraints. I illustrate here how the temperature may be quickly estimated in such a case.

The nightside temperature may be estimated by assuming the dominant source of heating is from the incident stellar flux. In this case, the only unknown factors affecting the nightside temperature are the re-distribution of heat factor, $\mathfrak{F}$ and the Bond albedo of the planet, $\mathfrak{A}_B$. First-order estimations of these values can be made based upon empirical upper limits and measurements of other planets and atmospheric models. The following expression may be used as a first-order estimate of the planet's brightness temperature:

$$T_{P,\text{hemisphere}}(\lambda) = \left(T_*^4 \frac{R_*^2}{4a_P^2} \left[\mathfrak{F} \frac{(1-\mathfrak{A}_B)}{\epsilon} \right] + \frac{L_P^{\text{internal}}}{4\pi R_P^2 \sigma_B \epsilon} \right)^{1/4} \tag{5.9}$$

where T_* is the effective temperature of the host star, $\mathfrak{F}$ is the distribution of energy to the hemisphere in question, ϵ is the emissivity of the planet ($0 \leq \epsilon \leq 1$), L_P^{internal} is the luminosity of the planet from internal heat generation (e.g. tidal heating, radioactivity), σ_B is the Stefan–Boltzmann constant.

For this calculation, L_P^{internal} is generally assumed to be negligible unless large tidal forces are expected as a result of a highly eccentric orbit, for example. All of the other quantities are typically measured except for $\mathfrak{F}$ and $\mathfrak{A}_B$. Choices for these values may come from atmospheric models or experience with other exoplanets.

5.2.4 Comparison to Other Effects

5.2.4.1 Starspots

Starspots have been observed within the transit events in several cases; e.g. Pont et al. [24], Dittmann et al. [25]. They typically have been observed to have a radius of less than or equal to a few Earth radii and are estimated to have temperatures from 100 to 1,000 K cooler than the rest of the stellar surface. When a planet passes over a starspot, it results in an increase in relative flux within the transit signal which is easily identified. If one assumes that the only starspot is the starspot which has been crossed, then accounting for the effect is quite trivial and may be incorporated in the light curve modelling.

What is much more troublesome are out-of-transit starspots for which one has no direct evidence. The presence of out-of-transit starspots will cause the transit depth to appear larger, in general. As a typical example, Czesla et al. [26] estimate that the effect can cause underestimations of the planetary radius by a fraction of $\sim$3 % for CoRoT-2b, which is a 1.6% change in the transit depth. This effect is larger than the nightside pollution effect by an order of magnitude.

However, these effects can only be present for spotty stars whereas nightside pollution simply requires a hot planet. Also, the spot coverage of a stellar surface varies periodically giving rise to an ultimately regular pattern which may therefore be corrected for. In contrast, the nightside pollution effect is not periodic: it is a constant offset in a single direction. Further, prior information such as a phase curve or a dayside eclipse places strong constraints on B_P and concordantly the estimation of this parameter is not an issue. Therefore, even for spotty stars, no-one would propose ignoring the effects of blending induced by a nearby companion star and so it can be seen that negating the self-blend of the planet would also be folly.

5.2.4.2 Temporal Variability of F_P and F_*

If the stellar flux or the nightside emission of the planet experiences temporal variations, then one would expect δ_{obs} to also change over time. I will here estimate the magnitude of this effect. The expected changes in nightside emission has not been studied in as much detail as that for the dayside, but one expects the magnitude of the variations to be very similar, based upon an energy budget argument. Rauscher et al. [27] have used shallow-layer circulation models to estimate variations in the dayside emission at the 1–10% level. This is consistent with the observations of HD 189733b's dayside flux by Agol et al. [23], who measure variations in the dayside below 10%. Typical stellar flux changes are at the $\lesssim 1\%$ level and so the ratio (F_P/F_*) is more likely to vary due to the planet than the star.

For the case of HD 189733b, I later show that $B_P = 1.002571$ at 8.0 μm. If F_P increased by $\pm 10\%$, this would correspond to $B_P = 1.002828$ causing the transit depth to vary by 6 ppm. This would be around an order of magnitude below *Spitzer*'s sensitivity but could be potentially close to a 1-σ effect for JWST. Nevertheless, the effect is sufficiently small that it is unlikely to be significant in most cases.

5.2.4.3 Limb Darkening

The nightside pollution effect is generally only relevant for hot-Jupiters at infrared wavelengths. As one moves towards the 10–30 μm wavelength range, the effects of limb darkening become negligible. The curvature of the transit trough is essentially flat. However, the limb of the star will possess a more complicated profile [28, 29] and this could potentially introduce errors into the fitting procedure. It is generally prudent to include even the very weak limb darkening effects when modelling such transits.

As a result, (5.6) and (5.7) should not be used to attempt to correct light curves which were fitted without accounting for nightside pollution. They do, however, offer a useful approximate quantification as to the direction and magnitude of any errors. A comparison between the predictions of (5.6) and (5.7) and the exact limb-darkened nightside-polluted light curve fits is given later in Sect. 5.2.5 for the example of HD 189733b. A discussion of this method required to produce this exact modelling is given in Sect. 5.2.3.

5.2.4.4 Significance at Visible Wavelengths

I briefly consider the value of including the nightside pollution effect at visible wavelengths, in particular for the *Kepler Mission*. Borucki et al. [30] recently reported visible-wavelength photometry for HAT-P-7b which exhibits a combination of ellipsoidal variations and a phase curve [22], as well as an occultation of depth (130 ± 11) ppm. HAT-P-7b is one of the very hottest transiting exoplanets discovered, so it offers a useful upper-limit example. For the purposes of nightside pollution, the maximum possible effect would occur if eclipse was both due to thermal emission alone and efficient day–night circulation. In this hypothetical scenario which maximizes the nightside pollution, one would have $B_P = 1.00013$. In the case of HAT-P-7b, the transit depth was reported to be $\delta_{\mathrm{obs}} = (6056 \pm 47)$ ppm by Welsh et al. [22], implying that the undiluted transit depth is larger by 0.79 ppm, or 0.017-σ. Therefore, as expected, visible wavelength transits will not, in general, be significantly affected by nightside pollution for even *Kepler* class photometry.

5.2.5 Applied Example: HD 189733b

5.2.5.1 *Spitzer* IRAC 8.0 μm Measurement

I will here provide an example of the empirical method of compensating for nightside pollution. I used the corrected data of HD 189733b's phase curve at 8.0 μm, as measured by Knutson et al. [20] (obtained through personal communication). I applied a median-stack smoothing function to the light curve with a one-minute window in order to identify the eclipse contact points. I find the flux of the star by taking the mean of fluxes between the second and third contact points during the occultation, weighting each point by the reported error. The standard deviation within this region is divided by the square root of the number of data points to provide the error on the mean. All fluxes are then divided by the derived stellar flux and the error on each flux stamp is propagated through, incorporating the error on the stellar flux estimate.

In order to determine the nightside flux, which is not the same as the minimum flux, I adopt a baseline defined as 30 min before first contact and 30 min after fourth contact and find a mean of $B_P = 1.002571 \pm 0.000048$. The average RMS in this baseline is 0.65 mmag/min. If it were possible for the nightside of the planet to induce an occultation, as the dayside does, one would measure an occultation nightside depth of $(0.256 \pm 0.023)\%$, whereas Charbonneau et al. [31] report a dayside occultation depth of $(0.391 \pm 0.022)\%$.

As discussed in Sect. 5.2.3, ellipsoidal variations can also be responsible for out-of-transit flux variations and can be potentially confused with the phase curve. For HD 189733b, I use Eq. (1) of Pfahl et al. [32] to estimate an ellipsoidal variation amplitude of 2.2 ppm. Given that the phase curve exhibits a variation 1,350 ppm amplitude, ellipsoidal variations can be neglected for the rest of this analysis.

I now produce two fits of the light curve: (1) no blending factor (2) blending factor B_{night} included. Each light curve is fitted independently assuming a fixed period of $P_P = 2.2185733$ days and zero orbital eccentricity.[4] The results of the fits are displayed in Table 5.1.

For the fitting, I used a Markov Chain Monte Carlo (MCMC) algorithm which employs the limb darkening model of Mandel and Agol [1] and I chose the light curve fitting parameter set: $\{\tau_T, p^2, \widetilde{T}^{\mathrm{one}}, b_{P,T}, OOT\}$. I used 125,000 trials with the first 25,000 discarded for burn-in. Employing the code of I. Ribas, a Kurucz (2006) style atmosphere is used to interpolate the four non-linear limb darkening coefficients [33], following the same methodology of Beaulieu et al. [34], giving us $c_1 = 0.2790207$, $c_2 = -0.1506885$, $c_3 = 0.0779481$ and $c_4 = -0.0087653$.

[4] This is done for simplicity, as our goal is merely to compare corrected versus uncorrected photometry, not re-derive more accurate system parameters.

I use the same local baseline as defined earlier, constituting 22,382 data points and assume a circular orbit.[5] At this stage, no outliers have yet been rejected but I proceed to fit the unbinned light curve. I take the best-fit light curve and subtract it from the data to obtain the residues and then look for outlier points. I then use the median-absolute-deviation (MAD) [35] to provide a robust estimate of the standard deviation of the data, as this parameter is highly resistant to outliers, and find MAD $= 2.92062 \times 10^{-3}$. Since there are 22,382 points, then the maximum expectant departure from a normal distribution is 4.08 standard deviations. Any points above this level are rejected, where the evaluation of the standard deviation comes from the MAD value multiplied by 1.4826, as appropriate for a normal distribution.[6] This procedure rejects any points with a residual deviation greater or equal to 0.0176749, corresponding to 10 points.[7]

The new light curve is then refitted in the normal way and I present the best-fit value in Table 5.1 find the uncorrected light curve has a depth of $\delta_{\rm obs} = 2.3824 \pm 0.0061\,\%$. For comparison, Knutson et al. [36] report the fitted 8.0 μm depth to be $2.387 \pm 0.006\,\%$, which is consistent with this value.[8] Applying the correction due to nightside pollution, I find the undiluted transit depth to be $\delta = 2.3884 \pm 0.0061\ \%$ meaning that the depth has increased by 60 ppm corresponding to 1-σ. From this example, it is clear that negating an effect which systematically biases transit depth measurements by $\sim$1-σ would be imprudent.

Using (5.6), to first-order in $(B_x - 1)$, I estimate that the impact parameter should be overestimated by 6.9×10^{-4}. The fits reveal a very similar value of 6.0×10^{-4}. Similarly, for (a_P/R_*), (5.7) predicts an underestimation of 3.4×10^{-3} whereas the light curve fit finds 6.2×10^{-3}. As expected, the effect of a blend is less pronounced on the other parameters.

Based on the difference in collecting area, it is expected JWST will achieve a precision $\sim$6.6 times greater than *Spitzer*, suggesting this nightside pollution effect will become significant at the ~ 5–10σ level for future infrared transit observations of hot-Jupiters. Additionally, the binning of multiple *Spitzer* transits would raise the significance of the effect. For example, Agol et al. [23] reported seven 8.0 μm transits of HD 189733b which, if globally fitted would increase the significance of the nightside pollution effect to 2.6-σ. Such a large effect cannot be justifiably neglected.

5.2.5.2 *Spitzer* MIPS 24 μm

Knutson et al. [36] measured the phase curve of HD 189733b with the MIPS instrument onboard *Spitzer* about a year after the observations of the 8.0 μm phase curve

[5] A circular orbit is used for the purposes of fair comparison with (5.6), and justified by the fact the goal here is not to produce refined parameters, merely illustrate the significance of nightside pollution.

[6] Although strictly a Poisson distribution, for 22,382 data points, the distribution is very well approximated by a Gaussian

[7] Note that the data has already been cleaned of outlier measurements, which is why the MAD rejection criteria only identifies 10 outliers from 22,382 points.

[8] The slight difference likely comes from a different limb darkening treatment.

for the same system. Using the original normalized-to-stellar-flux unbinned data (personal correspondence with H. Knutson), I took the mean of data points $\simeq 1$ h either side of the transit event, which exhibit an RMS of 2.1 mmag/min. I combined the two baseline estimates to calculate a nightside relative flux of $B_P = 1.00438 \pm 0.00025$.

Knutson et al. [36] reported a 24 μm transit depth of (2.396 ± 0.027) % and this re-analysis of the data yields $\delta_{\mathrm{obs}} = (2.398 \pm 0.019)$ %, where the fit has been performed using the same methodology as for 8.0 μm, except that I assume no limb darkening at 24 μm. As before, I apply the correction due to nightside pollution and estimate a new 24 μm transit depth of $\delta = (2.409 \pm 0.020)$ %, which increases the depth by $\sim$0.5-σ. Despite the absolute effect being larger than that at 8.0 μm, the difference is fewer standard deviations away due to the much poorer signal-to-noise of the transit event itself.

5.2.6 Pollution of the Transmission Spectrum

In the standard theory of transmission spectroscopy, planetary nightside emission is assumed to be negligible and thus disregarded (e.g. as explicitly stated in the foundational theory of Brown [37]). However, it has been shown here that the effect noticeably changes the transit depth for high quality photometry. Essentially, it is posited here that the "traditional" transmission spectrum is in fact a combination of transmission through the terminator and the self-blending caused by emission from the nightside.

One subtle point is that the nightside pollution effect is not something which can be accounted for in the modelling of the transmission spectrum. It is generally useful to think of the nightside pollution effect as an astrophysical blend which happens to be related to the planetary properties. A transmission spectrum is typically found by fitting a transit light curve at multiple wavelengths and then fitting a spectrum through the retrieved transit depths which models the planetary atmosphere. These routines usually make use of radiative transfer, chemical equilibrium, molecular line lists, etc to estimate the opacity of the atmosphere at each wavelength. However, attempting to increase the sophistication of these models would not accurately account for the self-blending scenario. Recall that each transit depth is obtained by fitting an eclipse model through the light curve time series. Critically, it is at this stage where blending needs to be accounted for. The transit signal plus blend should be modelled as such from the outset due to the subtle, and quite intricate, inter-dependencies between $b_{P,T}$, (a_P/R_*), p^2 and the limb darkening. Thus it can be seen that attempting to incorporate the effects later on is far more challenging and completely unnecessary than simply fitting each transit light curve with a physically accurate model in the first place.

In this section, I will estimate how different an exoplanet's transmission spectrum would appear with and without nightside pollution. The spectra generated in this section were computed by Giovanna Tinetti, who was the co-author on the paper

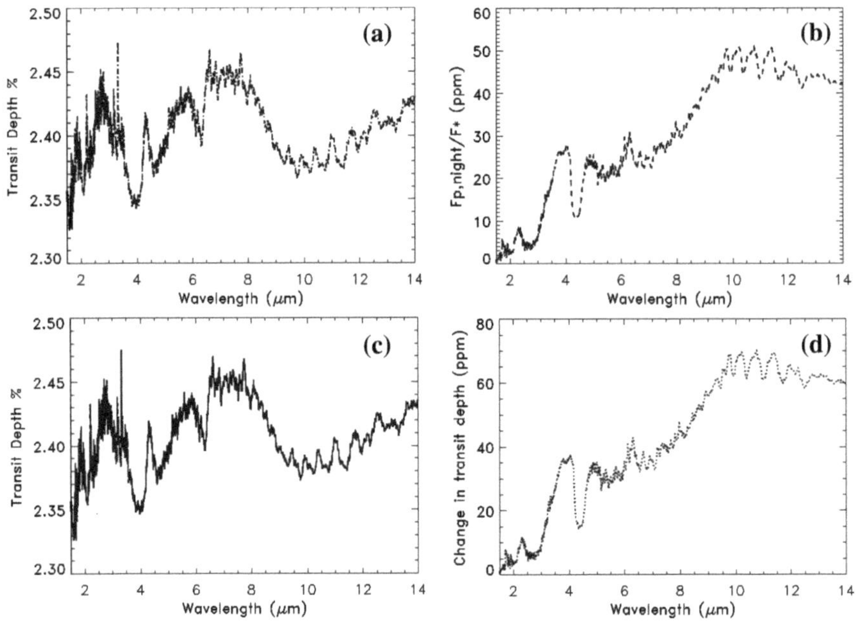

Fig. 5.2 *Top left* **a** Transmission spectrum of a hypothetical exoplanet similar to HD 189733b, generated considering the transmission through the terminator only. *Top right* **b** Emission spectrum from the nightside of the hypothetical planet. *Bottom left* **c** Transmission spectrum of the planet incorporating the pollution of the nightside emission. *Bottom right* **d** Residual between two transmission spectra. We conclude that not accounting for nightside emission would result in a 60–80 ppm error in the transit depth

on which this thesis chapter is based [2]. All other parts of this paper including the original concept, derivation and light curve analyses were conducted by myself.

In order to evaluate the magnitude of the spectral pollution, we will here consider a planet of similar type to HD 189733b. It is important to stress that the effects of nightside pollution will vary from case to case and the example we give here is indeed just for one example which is somewhat typical for an observed hot-Jupiter. Therefore the results here are only for a hypothetical, but typical, example.

The real question we need to answer is how much does a planetary transmission spectra change due to nightside self-blending? We therefore need to generate two versions of the planetary transmission spectra, one including (Fig. 5.2a) and one excluding the effects of nightside pollution (Fig. 5.2c), and then take the difference between the two (Fig. 5.2d).

For a description of the models used to generate the spectra, details may be found in Tinetti et al. [13] for the transmission spectrum and Swain et al. [15] for the emission. Planet and star properties are set to be that of the HD 189733 system. The model contains water, carbon dioxide and methane to provide us with the effects of molecular species on nightside pollution. No carbon monoxide or hazes/clouds are included in our example. We note that the transmission and emission models are good fits to the current available spectroscopy/photometry data of HD 189733b in the NIR/MIR [15]. The effects of water absorption are quantified with the BT2 water line list [38], which characterises water absorption at the range of temperatures probed in HD 189733b. Methane was simulated by using a combination of HITRAN 2008 [39] and PNNL data-lists. Carbon dioxide absorption coefficients were estimated with a combination of HITEMP and CDSD-1000 (Carbon Dioxide Spectroscopic Databank version for high temperature applications; [40]). The continuum was computed using $H_2 - H_2$ absorption data [41].

Generated spectra are always plotted in terms of the quantities determined with the lowest measurement uncertainty, namely $(R_P/R_*)^2$ and (F_P/F_*), for the transits and occultations respectively. Transmission spectra which are plotted in units of R_P (e.g. as done by Fortney et al. [42]) will cause the measurement uncertainties to be much larger since the error on R_* must necessarily be propagated in such a recipe. In fact, the measurement uncertainties on a spectra plotted in units of R_P will be dominated by the error on R_* since this property is typically determined to much lower precision. Consequently, statistically significant molecular features would be overwhelmed by the artificially large error bars.

Using the model described above, we first compute the transit depth from transmission absorption effects only (i.e. excluding nightside pollution) as visible in Fig. 5.2a. We then generate the dayside emission spectra for the same planet and make the assumption that the dayside and nightside emission spectra are identical (Fig. 5.2b). This assumption is unlikely to be true for the exact case of HD 189733b and really constitutes an upper limit, but we again stress that we are here only considering a planet similar to that of HD 189733b and thus we are free to make this assumption for our hypothetical example.

We combine the nightside emission and transmission spectra to produce a transmission spectra which includes the effects of the nightside, as seen in Fig. 5.2c. We then take the difference between the corrected spectrum and the one which excludes nightside emission. The resultant residual spectrum is plotted in Fig. 5.2d. The residual spectrum reveals nightside emission affects the transmission spectra at the level of 6×10^{-5} for $\lambda \geq 10$ μm and very closely matches the behaviour of the emission features, as expected. As we saw earlier, the magnitude of the effect is equal to the typical measurement uncertainty for a target like HD 189733b with *Spitzer*. This supports our hypothesis that the nightside pollution effect is a $\sim$ 1-σ effect for 8.0 μm *Spitzer* photometry.

5.2.7 Extrapolating the Nightside Correction for HD 189733b

Only two measurements exist for the nightside flux for HD 189733b (or indeed any other exoplanet) at 8.0 and 24 μm, but several other primary transit light curves exist in the mid-infrared wavelengths. Beaulieu et al. [43] presented 3.6 and 5.8 μm measurements and Désert et al. [44] obtained photometry at 4.5 and 8.0 μm. I will here estimate the nightside effect on the 3.6, 4.5 and 5.8 μm channels. Currently, only two data points exist and this does not warrant modelling the nightside emission spectrum in any more complexity than that of a blackbody, as a first-order approximation. Should more measurements become available in the future, molecular species could be included to improve the accuracy of our model.

I first assume that the star is blackbody emitter with $T_* = (5040 \pm 50)$ K [45]. At this point is advantageous to consider only the emission per unit area from both the planet and the star, in order to avoid the effects of transmission through the planetary atmosphere. I therefore define the flux per unit area of each object using:

$$\mathcal{F}_P = F_P/(\pi R_P^2) \tag{5.10}$$

$$\mathcal{F}_* = F_*/(\pi R_*^2) \tag{5.11}$$

$$\mathcal{R}_P = \frac{\mathcal{F}_P}{\mathcal{F}_*} = \frac{F_P}{F_*}\delta \tag{5.12}$$

I then convert the B_P measurements for 8.0 and 24 μm into $\mathcal{R}_P$ by using the two values of the corrected ratio-of-radii shown in Tables 5.1 and 5.2 and propagating the uncertainties. Using the Planck function as a blackbody model for each body and the relevant *Spitzer* bandpass response functions, one may estimate $\mathcal{R}_{\text{night}}$ for any given value of $T_{P,\text{night}}$. I allow this temperature to vary from 700 to 1,700 K in 1 K steps and numerically integrate the bandpasses to find χ^2 at each temperature, which I define as:

$$\chi^2 = \left(\frac{\mathcal{F}_{\text{obs}} - \mathcal{F}_{\text{calc}}}{\Delta(\mathcal{F}_{\text{obs}})}\right)^2\Bigg|_{8.0\,\mu\text{m}} + \left(\frac{\mathcal{F}_{\text{obs}} - \mathcal{F}_{\text{calc}}}{\Delta(\mathcal{F}_{\text{obs}})}\right)^2\Bigg|_{24\,\mu\text{m}} + \left(\frac{T_{*,obs} - T_{*,calc}}{\Delta(T_*)}\right)^2 \tag{5.13}$$

where I additionally define $T_{*,\text{calc}}$ as the temperature used in the integration and $T_{*,\text{obs}}$ as being equal to the value determined by Torres et al. [45]. Note that I do not fit for T_* but do allow the value to float in order to correctly estimate the uncertainty of $T_{P,\text{night}}$. The final analysis reveals a best-fit planetary nightside temperature of $T_{P,\text{night}} = (1,148 \pm 32)$ K with $\chi^2 = 1.12$, suggesting the blackbody model gives a satisfactory fit for these two measurements. Note that not accounting for the response function of the instruments would yield a erroneous result of $T_{P,\text{night}} = 1,120$ K.

Using our derived planetary temperature, one may now use the blackbody function to extrapolate $\mathcal{R}_{\text{night}}$ to other wavelengths and thus the nightside corrected transit depths for 3.6, 4.5 and 5.8 μm. The undiluted transit depth will be given by Equation (5.14) and the results of this analysis are summarized in Table 5.3 and Fig. 5.3.

$$\delta = \frac{\delta_{\text{obs}}}{1 - \mathcal{R}_P\delta_{\text{obs}}} \tag{5.14}$$

The maximal deviation occurs for 8.0 μm and is less than 1-σ for all other wavelengths. Consequently, the deduction of which molecules are evident from the spectrum of HD 189733b will not significantly affected by the nightside effect, but derived abundances will change.

Table 5.3 Using a fitted blackbody function for the nightside emission of HD 189733b, we calculate the nightside corrections to *Spitzer* channels for which no phase curve information currently exists

Channel	Observed depth δ_{obs} (%)	$\mathcal{R}_P$	Corrected depth δ, (%)	$\frac{\delta-\delta_{obs}}{\sigma_\delta}$
Measured				
8.0 μm	2.3824 ± 0.0060	0.1076 ± 0.0020	2.3884 ± 0.0061	1.0
24.0 μm	2.398 ± 0.019	0.1818 ± 0.0105	2.4085 ± 0.020	0.5
Extrapolated				
3.6 μm	2.356 ± 0.019	$0.0365^{\dagger}$	$2.358 \pm 0.019^{\dagger}$	≤ 0.1
4.5 μm	2.424 ± 0.010	$0.0570^{\dagger}$	$2.427 \pm 0.010^{\dagger}$	≤ 0.4
5.8 μm	2.436 ± 0.020	$0.0800^{\dagger}$	$2.441 \pm 0.020^{\dagger}$	≤ 0.25

Values with a † superscript cannot have their uncertainties estimated since they are extrapolated parameters. 8.0 μm data comes from Knutson et al. [20], 24.0 μm from Knutson et al. [36], 3.6 and 5.8 μm from Beaulieu et al. [33] and 4.5 μm from Désert et al. [41]

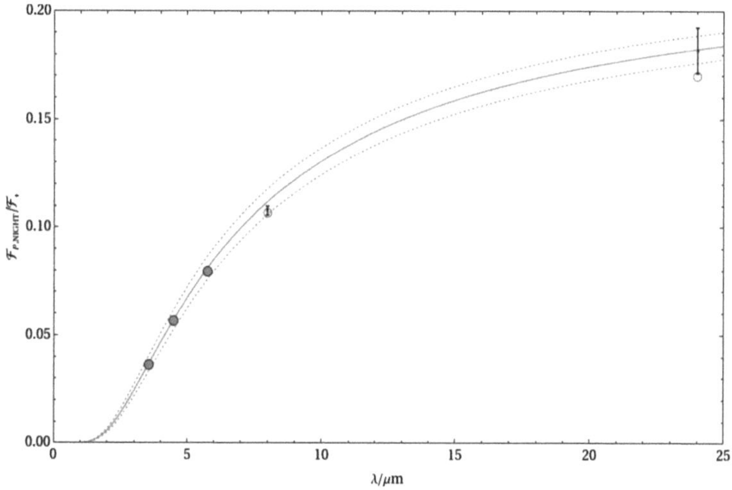

Fig. 5.3 Flux per unit area (i.e. not total flux) of the planetary *nightside* emission divided by that of the star, plotted as a function of wavelength. Using the two measurements at 8.0 and 24.0 μm (*black dots* with error bars), we fit a blackbody curve through the points (*gray lines*) with a nightside planetary temperature of $T = 1148 \pm 32$ K *Open circles* represent the integrated blackbody function across the instrument bandpasses. *Filled circles* represent the same but extrapolated to the other IRAC wavelengths, which allow us to conclude the nightside effect will be much less at lower wavelengths

5.2.8 Conclusions

Both transit photometry and transmission spectroscopy are expected to be affected by self-blending of the planet's nightside emission at infrared wavelengths. This self-blending, or "nightside pollution", causes a dilution of transit depths and invalidates assumptions A4 and A5. However, I have shown here how either high resolution imaging, for extrinsic blend sources, or phase curve observations, for self-blends, can be used to determine the extra flux contribution and thus compensate for the effect. In a worked example, it is shown that HD 189733b's 8.0 μm *Spitzer* transit light curve experiences a 1-σ bias if this effect is not accounted for. With A4/A5 now addressed, the outstanding assumption which is frequently broken in A10, which I move onto next in Sect. 5.3.

5.3 Binning is Sinning

5.3.1 Introduction

The final assumption to be considered is A10—"The change in flux over a single integration is much smaller than the flux measurement uncertainty". This essentially assumes a very short integration time but the quantification of just how short this really is has not been provided. This may be evaluated by breaking the assumption and then computing when the consequences surpass the noise. Assumption A10 was never stated in Mandel and Agol [1] or Seager and Mallén-Ornelas [9], but it can be seen to be implicit in both pioneering papers. Much like how nightside pollution was never conceived until precise infrared measurements became available with *Spitzer*, the issue of A10 was never considered until a space-based telescope combining long-integrations with extremely high precision photometry was launched i.e. *Kepler*.

The *Kepler Mission* surveys more than 150,000 stars for transiting planets and thus produces large amounts of data. Due to telemetry limits, almost all of these measurements are performed in long-cadence (LC) mode to preserve bandwidth. The LC mode utilizes 30 min integrations and so one has the combination of integration times of comparable size to the transit features plus high precision photometry. This combination maximizes the conditions under which A10 is broken. In this section, I will consider the consequences of this invalidation, and how one can compensate for the effect.

5.3.2 The Effects of Finite Integration Time

5.3.2.1 Ingress/Egress Durations

As discussed in Sect. 3.3.3 there are four critical contact points which define the transit light curve morphology, which represent the points where the time derivative is discontinuous. Physically speaking, contact points I and IV occur when the sky-projected planet-star separation is equal to the stellar radius plus the planetary radius. Contact points II and III occur when this parameter equals the stellar radius minus the planetary radius. Defining W as the mean of the durations between the first-to-fourth and second-to-third contacts, one may write:

$$t_{\text{I}} = \tau_T - W/2 - T_{12}/2 \tag{5.15}$$

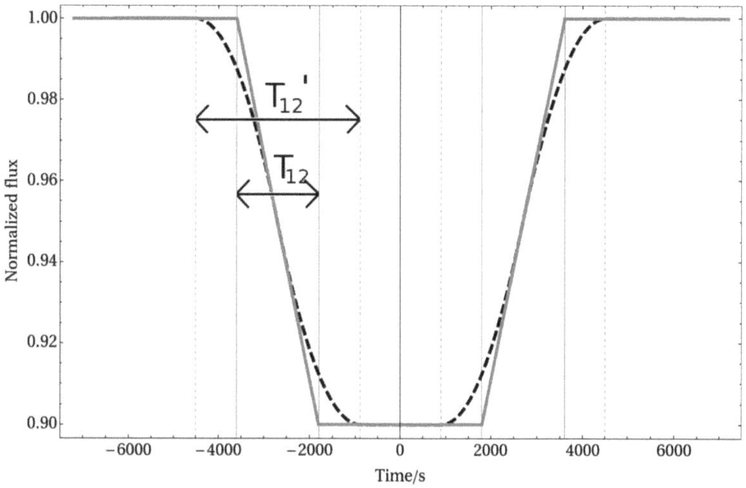

Fig. 5.4 A trapezoid approximated light curve with a 1 h flat-bottom duration and 30 min ingress/egress duration, T_{12}, is shown in solid. The *dashed line* shows the light curve morphology for an integration time of 30 min. The apparent ingress/egress duration, T'_{12}, can be seen to have doubled purely as a consequence of the integration time

$$t_{\text{II}} = \tau_T - W/2 + T_{12}/2 \tag{5.16}$$

$$t_{\text{III}} = \tau_T + W/2 - T_{12}/2 \tag{5.17}$$

$$t_{\text{IV}} = \tau_T + W/2 + T_{12}/2 \tag{5.18}$$

The principal effect of finite integration time is to smear out the light curve into a broader shape (see Fig. 5.4). The apparent ingress and egress durations will increase and additional curvature will be introduced into the light curve wings. The ingress/egress stretching can be considered in terms of the apparent positions of the contact points being temporally shifted from their true value. The magnitude of this time shift is dependent on the relative phase difference between the sampling and the transit signal. If one assumes that a large number of transits observed with LC photometry are folded about the orbital period, as is typical in transit detection, then the effect becomes much more predictable with the deviation averaging out to $\mathcal{I}/2$.

Under these conditions, contact points I and IV will appear to move outwards from τ_T by one half of the integration time each, $\mathcal{I}/2$. Conversely, contact points II and III will appear to move inwards by the same amount. Let us define the apparent contact points as t'_x:

$$t'_{\text{I}} = t_{\text{I}} - \mathcal{I}/2 \tag{5.19}$$

$$t'_{\mathrm{II}} = t_{\mathrm{II}} + \mathcal{I}/2 \tag{5.20}$$

$$t'_{\mathrm{III}} = t_{\mathrm{III}} - \mathcal{I}/2 \tag{5.21}$$

$$t'_{\mathrm{IV}} = t_{\mathrm{IV}} + \mathcal{I}/2 \tag{5.22}$$

These shifted apparent contact points will cause the apparent transit durations T'_{14} and T'_{23} to differ from the true values. Specifically, an observer would find $T'_{14} = (T_{14} + \mathcal{I})$ and $T'_{23} = (T_{23} - \mathcal{I})$. If the observer was unaware that this smearing had occurred, they would naively use Eq. (4.81) (or alternatively the Seager and Mallén-Ornelas [9] equations if one assumes $e_P = 0$) to retrieve erroneous values of $b_{P,T}$, (a_P/R_*), i_P and ρ_*. I denote these erroneous retrieved parameters with the superscript "instant", in reference to the assumption of instantaneous integration times. They may be computed using the same method I used in Sect. 5.2, and utilizing (4.81) I find:

$$[b_{P,T}^{\mathrm{instant}}]^2 = \frac{\left[(1+p)^2 \dfrac{\sin^2\left[\dfrac{\sqrt{1-e_P^2}\mathcal{I}\pi}{P_P \varrho_{P,T}^2} - \arcsin\left(\dfrac{\sqrt{(1-p)^2-b_{P,T}^2}}{(a_P/R_*)\varrho_{P,T}\sin i_P}\right)\right]}{\sin^2\left[\dfrac{\sqrt{1-e_P^2}\mathcal{I}\pi}{P_P \varrho_{P,T}^2} + \arcsin\left(\dfrac{\sqrt{(1+p)^2-b_{P,T}^2}}{(a_P/R_*)\varrho_{P,T}\sin i_P}\right)\right]} - (1-p)^2\right]}{\left[\dfrac{\sin^2\left[\dfrac{\sqrt{1-e_P^2}\mathcal{I}\pi}{P_P \varrho_{P,T}^2} - \arcsin\left(\dfrac{\sqrt{(1-p)^2-b_{P,T}^2}}{(a_P/R_*)\varrho_{P,T}\sin i_P}\right)\right]}{\sin^2\left[\dfrac{\sqrt{1-e_P^2}\mathcal{I}\pi}{P_P \varrho_{P,T}^2} + \arcsin\left(\dfrac{\sqrt{(1+p)^2-b_{P,T}^2}}{(a_P/R_*)\varrho_{P,T}\sin i_P}\right)\right]} - 1\right]} \tag{5.23}$$

$$[(a_P/R_*)^{\mathrm{instant}}]^2 = \frac{\dfrac{\left[(1+p)^2 \sin^2\left(\dfrac{\sqrt{1-e_P^2}\pi\mathcal{I}}{\varrho_{P,T}^2 P_P} - \arcsin\left(\dfrac{\sqrt{(1-p)^2-b_{P,T}^2}}{(a_P/R_*)\varrho_{P,T}\sin i_P}\right)\right) - 4p\right]}{\sin^2\left(\dfrac{\sqrt{1-e^2}\pi\mathcal{I}}{\varrho_c^2 P} + \arcsin\left(\dfrac{\sqrt{(1+p)^2-b_{P,T}^2}}{(a_P/R_*)\varrho_c\sin i_P}\right)\right)} - (1-p)^2}{\varrho_{P,T}^2 \dfrac{\sin^2\left(\dfrac{\sqrt{1-e_P^2}\pi\mathcal{I}}{\varrho_{P,T}^2 P_P} - \arcsin\left(\dfrac{\sqrt{(1-p)^2-b_{P,T}^2}}{(a_P/R_*)\varrho_{P,T}\sin i_P}\right)\right)}{\sin^2\left(\dfrac{\sqrt{1-e_P^2}\pi\mathcal{I}}{\varrho_{P,T}^2 P_P} + \arcsin\left(\dfrac{\sqrt{(1+p)^2-b_{P,T}^2}}{(a_P/R_*)\varrho_{P,T}\sin i_P}\right)\right)} - \varrho_{P,T}^2} \tag{5.24}$$

$$\rho_*^{\mathrm{instant}} \simeq \frac{3\pi}{G P_P^2}[(a_P/R_*)^{\mathrm{instant}}]^3 \tag{5.25}$$

Setting $\mathcal{I} = 0$ returns the original results as expected. Unfortunately, these equations are somewhat overly complex for one to draw any physical intuition. To proceed, I will consider a typical case example by using the system parameters from one of the *Kepler* planets, since these are discovered using LC data. The following example is for the assumption of zero limb darkening, which is a very poor one for the *Kepler* bandpass. The effects of limb darkening will be discussed later.

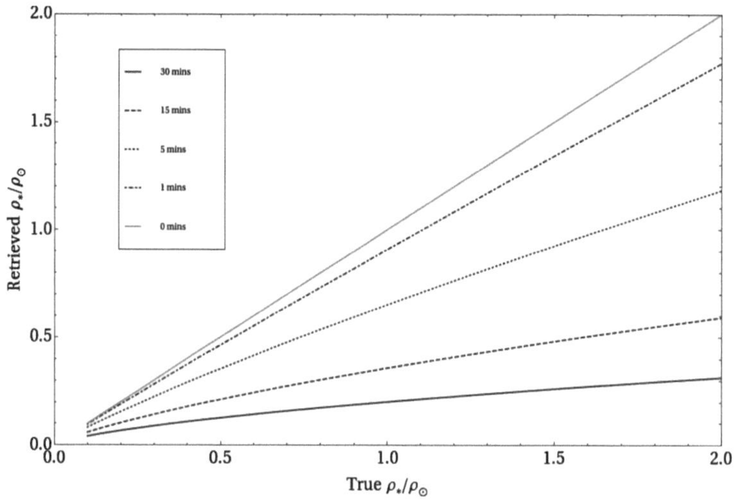

Fig. 5.5 As an example, I use the system parameters of Kepler-5b to show the effect on the retrieved stellar density as a result of long integration times, in the case of no limb darkening. The 1 min line appears to produce results within the typical uncertainties of the derived stellar density

In Fig. 5.5, I plot the retrieved stellar density, $\rho_*^{\mathrm{instant}}$, as a function of the true stellar density, ρ_*. All other parameters are fixed to be that of Kepler-5b, as reported by Koch et al. [46]. The effect can be seen to be highly significant, causing the retrieved stellar density to be underestimated by a factor which borders on an order-of-magnitude. This scale of underestimation is sufficient to completely reject some planetary candidates as unphysical. However, note that in reality the underestimation of ρ_* will not be this severe due to countering effects of limb darkening suppression discussed later.

Looking at Fig. 5.5 again, let us explore the physical reasons for this underestimation. I begin this line of thought by considering the impact parameter, $b_{P,T}$. If the ingress/egress duration is elongated, what does one expect to happen to the derived impact parameter? Consider two extreme cases. When a planet transits with a very low impact parameter, one has essentially an equatorial transit. This means the vector describing the sky-projected planetary velocity is nearly perpendicular to the stellar limb. As a result, the planet crosses the limb expediently. In contrast, for a near-grazing transit, the mutual angle between the stellar limb and the sky-projected planetary velocity vector has become more acute, which has the effect of making the limb-crossing time much longer. Therefore, stretching the ingress/egress duration causes $b_{P,T}^{\mathrm{instant}} > b_{P,T}$.

Having established this point, consider the effect on (a_P/R_*). The simplest way to understand the effect on this parameter is to appreciate that $b_{P,T}$ and (a_P/R_*) exhibit an extremely strong negative correlation, as demonstrated by Carter et al. [47] and seen earlier in Table 4.1. So the act of increasing the ingress/egress duration will increase $b_{P,T}$ and therefore decrease (a_P/R_*). Finally, one knows that $\rho_* \propto$

$(a_P/R_*)^3$ meaning that decreasing (a_P/R_*) by a factor of 2 would decrease ρ_* by a factor of 8.

5.3.2.2 Limb Darkening Effects

For transit observations at visible wavelengths, limb darkening is quite pronounced producing a well-known curvature in the flat-bottom part of the light curve. For large integration times, the curvature is smeared out, producing a flatter transit trough morphology.

However, for stars exhibiting limb darkening, the boundary between the end of ingress and start of the trough is less obvious and can form an essentially continuous curve. As a result, an algorithm which is fitting this light curve may "borrow" some of the ingress portion to increase the net curvature in what the algorithm considers to be the trough. The reason an algorithm would do this is because usually the limb darkening parameters are known and often fixed and so the algorithm is trying to find a solution which produces the expected curvature a-priori.

By "borrowing" some of the ingress, the overall effect of limb darkening is to actually decrease the systematic overestimation of the ingress discussed in the last subsubsection. Therefore, LC data pushes $b_{P,T}^{\text{instant}}$ towards a more equatorial value due to limb darkening.

5.3.2.3 Two Countering Effects

In conclusion, finite integration times have two countering effects on $b_{P,T}^{\text{instant}}$ and thus $(a_P/R_*)^{\text{instant}}$ and ρ_*^{instant} as well. The ingress and egress smearing causes the ingress to appear larger, which occurs for more grazing transits. In contrast, the limb darkening borrows some of the ingress and thuds attenuates the previous effect, causing an apparently more equatorial transit. Whilst limb darkening is important, especially for transit surveys like *CoRoT* and *Kepler* which operate at visible wavelengths, the fundamental change in the transit morphology is sufficiently large that it will tend to dominate. This is because the amplitude of the curvature in the light curve trough is usually at least an order-of-magnitude less than the amplitude of the transit signal itself. Therefore, the consequences of smearing the overall transit signal will tend to dominate over the effects of limb darkening.

This general rule of thumb that $b_{P,T}^{\text{instant}}$ is overestimated will break-down when one has $T_{12} \gg \mathcal{I}$ but $T_{23} \sim \mathcal{O}[\mathcal{I}]$, which occurs for near-grazing transits. In such a case, the fractional change to the ingress durations is minimized but the change in the transit trough curvature is maximized.

To exactly calculate the net consequence of these two effects though, the integration time should be included when one generates the model light curves, rather than attempting ad hoc corrections post-analysis. This appears to be the only way to completely account for the effect in a reliable manner.

5.3.2.4 Consequences for the Transit Depth

I will briefly comment on the effect of finite integration time on the transit depth. Assuming no limb darkening is present and $T_{23} > \mathcal{I}$, the transit depth is completely unaffected by the long integration time. This has important consequences for occultations where the light curve is unaffected by stellar limb darkening.

For cases where limb darkening is present, the net effect on the retrieved p^{instant} will depend on whether $b_{P,T}^{\text{instant}} > b_{P,T}$ or $b_{P,T} < b_{P,T}^{\text{instant}}$. Additionally, the effect will be a function of what assumptions were used in the fitting algorithm (e.g. the fixing of various parameters), the true impact parameter and the limb darkening coefficients. Given the large number of correlated factors, predicting the effect of integration time becomes less reliable and one must fit the transit light curve with a model which accounts for integration time in the first place.

5.3.2.5 Observed Effects with *Kepler*

I point out that the effects of long integration times have already been observed by the *Kepler Mission*. Figure 4 of Gilliland et al. [48] shows the transit light curve of long-cadence and short-cadence data for the same planet, TreS-2b. The long-cadence light curve exhibits a broader shape with the apparent position of the contact points shifted by $\sim \mathcal{I}/2$, as predicted in this work. Notice also that the curvature in the transit trough, due to limb darkening, has also been attenuated.

5.3.3 Accurate Transit Light Curve Modelling

5.3.3.1 Analytic Integration

The critical problem I have outlined can be simply summarized by the following: *Don't fit an unbinned model to binned data*. The model usually used to generate a transit light curve is provided by Mandel and Agol [1], which includes the effects of stellar limb darkening. To generate the Mandel and Agol [1] light curve, one usually has a set of time stamps forming a time vector **t**. This time vector represents instantaneous moments rather than integrated time. The **t** vector is converted into a vector of instantaneous true anomalies, $\mathbf{f_P}$, by solving Kepler's Equation numerically (see Sect. 3.1.2). $\mathbf{f_P}$ is converted to a $\mathbf{S_{P*}}$ array, where $\mathbf{S_{P*}}$ is an array of the instantaneous sky-projected planet-star separations. Finally, the Mandel and Agol [1] equations provide us with **F**, where F is the instantaneous flux. The sequence of events may be summarized as:

$$\mathbf{t} \rightarrow \mathbf{f_P} \rightarrow \mathbf{S_{P*}} \rightarrow \mathbf{F(t)}$$

Now that the mechanism of generating of a transit light curve for instantaneous time stamps, $\mathbf{F(t)}$, has been established, I will consider what the transit lightcurve for integrated time stamps would be, $\check{\mathbf{F}}(\check{\mathbf{t}})$. In this case, the integrated flux of the ith data point would be given by:

$$\check{F}_i(\check{t}_i) = \frac{\int_{t=\check{t}_i-\mathcal{I}/2}^{\check{t}_i+\mathcal{I}/2} F(t)\,\mathrm{d}t}{\int_{t=\check{t}_i-\mathcal{I}/2}^{\check{t}_i+\mathcal{I}/2} \mathrm{d}t} \tag{5.26}$$

This equation suffers from the problem that F cannot be written as a function of t analytically, since such a solution would require a closed-form solution to Kepler's Equation, which is transcendental. Evaluating this expression for F as a function of S_{P*} is also not possible since one would find the following:

$$\int\limits_{t=\check{t}_i-\mathcal{I}/2}^{\check{t}_i+\mathcal{I}/2} F(t)\,\mathrm{d}t = \int\limits_{S_{P*}(\check{t}_i-\mathcal{I}/2)}^{S_{P*}(\check{t}_i+\mathcal{I}/2)} F(t(S_{P*}))\left[\frac{\mathrm{d}t}{\mathrm{d}S_{P*}}(S_{P*})\right]\mathrm{d}S_{P*} \tag{5.27}$$

Whilst $\mathrm{d}S_{P*}/\mathrm{d}t$ may be evaluated analytically through a chain rule expansion of $(\mathrm{d}S_{P*}/\mathrm{d}f_P) \times (\mathrm{d}f_P/\mathrm{d}t)$, the resultant expression will be as a function of f_P, rather than S_{P*}. If one knew $f_P(S_{P*})$, then one would be able to write out the integrand in a closed-form, but $f_P(S_{P*})$ can only be found by solving a bi-quartic equation, as was shown earlier in Sect. 4.3.1. Unfortunately, as was discussed, there is no currently proposed method to find which roots correspond to which orbital conjunction which makes a closed-form expression elusive.

The only remaining hope for a simple analytic expression would be to express the integral in terms of true or eccentric anomaly, which are inter-changeable. This would yield the following integral:

$$\int\limits_{t=\check{t}_i-\mathcal{I}/2}^{\check{t}_i+\mathcal{I}/2} F(t)\,\mathrm{d}t = \int\limits_{f_P(\check{t}_i-\mathcal{I}/2)}^{f_P(\check{t}_i+\mathcal{I}/2)} F(t(f_P))\frac{\mathrm{d}t}{\mathrm{d}f_P}(f_P)\,\mathrm{d}f_P \tag{5.28}$$

The integrand of this expression may be written out in a closed-form, using (3.11):

$$\int\limits_{t=\check{t}_i-\mathcal{I}/2}^{\check{t}_i+\mathcal{I}/2} F(t)\,\mathrm{d}t = \int\limits_{f_P(\check{t}_i-\mathcal{I}/2)}^{f_P(\check{t}_i+\mathcal{I}/2)} F(S_{P*}(f_P))\frac{P_P}{2\pi}\frac{(1-e_P^2)^{3/2}}{(1+e_P\cos f_P)^2}\,\mathrm{d}f_P \tag{5.29}$$

where $F(S_{P*})$ is given by Mandel and Agol [1] and $S_{P*}(f_P)$ is given in (3.19). The integral limits do not possess a closed-form solution since once again one must solve Kepler's Equation, but in principle the indefinite integral could be analytically evaluated and then the relevant limits applied after a subroutine provides numerical solutions to Kepler's Equation. I believe that this strategy would be the most computationally efficient since this would obviate the need for any numerical integration.

However, I was unable to find a solution for the indefinite integral for even a uniform-source case and will therefore focus the remainder of this discussion onto numerical techniques.

5.3.3.2 Numerical Integration

Having established the significant challenges regarding analytic integration, I now consider numerical integration techniques. The functions one needs to integrate over are in fact very well-behaved and well-approximated by compositions of polynomials and thus I anticipate that even a low-resolution numerical integration technique should provide satisfactory accuracy.

In this subsection, I first consider the merit of Simpson's Rule or other Newton–Cotes based methods. I aim to avoid using nested quadrature methods like the Gauss–Kronrod or Clenshaw–Curtis, as the number of integrations required is large and it is preferable to avoid nested methods. For the simplest case of Simpson's rule, the integrated flux is given by:

$$\check{F}_i(N=3) = \frac{F(t_i - \mathcal{I}/2) + 4F(t_i) + F(t_i + \mathcal{I}/2)}{6} \tag{5.30}$$

where N denotes the number of calls needed to the Mandel and Agol [1] code and essentially is a measure of the resolution of the numerical integration. This method may be extended to higher intervals by using Simpson's composite rule. Alternatively, one can extend to cubic, quartic, etc interpolations by using the Newton–Cotes formulas. Each time one evaluates $F(t)$, one requires another call to the Mandel and Agol [1] subroutine, and thus one wishes to minimize the number of calls, but maximize the accuracy of the employed technique.

Simpson's composite rule works by splitting up our integration range into $2m$ subintervals, therefore requiring $N = 2m + 1$ calls to the Mandel and Agol [1] code. The error on the composite Simpson's rule is given by:

$$\underset{\text{Comp. Simpson}}{\sigma_{\check{F}}(N)} = F^{(4)}(\epsilon)\frac{\mathcal{I}}{180}\left(\frac{\mathcal{I}}{N-1}\right)^4 \tag{5.31}$$

where ϵ lies in the range $t_i - \mathcal{I}/2 < \epsilon < t_i + \mathcal{I}/2$. In contrast, the Newton–Cotes formulas move through increasing orders by increasing the interpolation order. For the $N = 4$ case (which is the cubic interpolation scenario, known as Simpson's 3/8 rule), the equivalent errors between the two methods are:

$$\underset{\text{Newton–Cotes}}{\sigma_{\check{F}}(N=4)} = F^{(4)}(\epsilon)\frac{3}{80}\mathcal{I}^5 \tag{5.32}$$

$$\underset{\text{Comp. Simpson}}{\sigma_{\check{F}}(N=4)} = F^{(4)}(\epsilon)\frac{1}{2880}\mathcal{I}^5 \tag{5.33}$$

Thus for $N = 4$, Simpson's composite rule offers greater accuracy than the Newton–Cotes based equation. Moving through the higher orders in the Newton–Cotes family causes the error to have a functional dependence on $F^{(N)}$, i.e. the Nth differential of F. So for $N > 4$ it is not possible to give an exact comparison between the two methods since $F^{(N)}(t)$ is not known for any $N > 1$. Therefore, our only reliable comparison is for the $N = 4$ case, from which I conclude the composite Simpson's rule is superior in terms of accuracy versus computational requirement.

5.3.3.3 Error in Numerical Integration

I now consider what value of N should be used. There are essentially three segments of the light curve which exhibit curvature and thus would produce the maximum error in the numerical integrations, which employ linear piece-wise approximations.

- Curvature of the ingress/egress
- Curvature of the limb-darkened light curve trough
- Discontinuities at the contact points

The last of these is due to a discontinuous function and the former two are due to curvatures within continuous functions. I will treat these two different sources of "curvature" separately, although from the arguments made earlier, one expects the last of these effects to be the largest source of numerical error.

Ingress/Egress Curvature

The transit light curve has a depth δ and an ingress duration T_{12}. For most of the ingress, the curvature is close to zero and essentially mimics a linear slope. However, near the contact points, the slope rapidly changes to a flat line of zero gradient. Therefore, near the contact points, the ingress/egress morphology causes large amounts of curvature. These points will exhibit the largest numerical errors in using a technique like Simpson's composite rule.

A suitable choice of resolution can be made by increasing N until $\sigma_{\check{F}}|_{\max} \leq \sigma_{\check{F},\text{obs}}$, i.e. the calculation should produce a flux which has a maximum systematic error which is less than the observational uncertainty. One should set the resolution to a point where it provides satisfactory accuracy even at the point of highest numerical error, i.e. within the ingress/egress near the contact points.

Another approach would be to use an adaptive composite Simpson's rule, for example like that proposed by McKeeman (1962). However, it is preferable to avoid using adaptive routines since they would require a new adaptation for every single data point and fitting trial, which would be time consuming. The costs versus benefits of using such a method could warrant further investigation in the future. Instead, I choose to use the adaptation required for the most troublesome points, which have already been identified. The required interval size in each element of the Simpson's composition should be decreased until one reaches:

$$\left| \mathcal{S}\left(a, \frac{a+b}{2}\right) + \mathcal{S}\left(\frac{a+b}{2}, b\right) - \mathcal{S}(a, b) \right| /15 < \sigma_{\check{F},\mathrm{obs}} \tag{5.34}$$

where $\mathcal{S}(\alpha, \beta)$ is Simpson's rule evaluated over the interval α to β. In this case, the integral is over time and $a = t_I$ and $b = t_I + (\mathcal{I}_0/m)$, where $2m$ is the number of subintervals which the integral has been split into and $2m = N - 1$, where N is the required factor by which the number of calls to the Mandel and Agol [1] code increases. The reason for the subscript of 0 by the $\mathcal{I}$ term will be explained shortly. This requirement may be written as:

$$\frac{|\mathcal{S}(t_I, t_I + \frac{\mathcal{I}_0}{2m}) + \mathcal{S}(t_I + \frac{\mathcal{I}_0}{2m}, t_I + \frac{\mathcal{I}_0}{m}) - \mathcal{S}(t_I, t_I + \frac{\mathcal{I}_0}{m})|}{15} < \sigma_{\check{F},\mathrm{obs}} \tag{5.35}$$

$$\mathcal{S}(\alpha, \beta) = \left(\frac{\beta - \alpha}{6}\right)\left[F(\alpha) + 4F\left(\frac{\alpha+\beta}{2}\right) + F(\beta)\right] \tag{5.36}$$

In order to continue, one needs to evaluate *F(t)* in a closed-form, which cannot be achieved due to Kepler's Equation. However, there exists a special case where Kepler's Equation does yield an exact closed-form solution and this occurs for circular orbits since $\mathfrak{M}_P = E_P = f_P$. In such a case, one may use (4.33), which I remind the reader of here:

$$S_{P*}(t) = (a_P/R_*)\sqrt{\sin^2\left(\frac{2\pi t}{P_P}\right) + \cos^2 i_P \cos^2\left(\frac{2\pi t}{P_P}\right)} \tag{5.37}$$

The ingress/egress morphology is dominated by the expressions pertaining to a uniform source. Limb darkening does affect the ingress/egress curvature but this is much less than the amplitude of the uniform source transit signal. In the small-planet limit, Mandel and Agol [1] provided the following approximation for the ingress/egress flux:

$$F(\mathcal{X}) = 1 + \mathcal{X}\sqrt{p^2 - \mathcal{X}^2} - p^2 \arccos\left[\frac{\mathcal{X}}{p}\right] \tag{5.38}$$

where I have defined $S_{P*} = 1 + \mathcal{X}$ and it is understood that $-p < \mathcal{X} < p$ for the ingress/egress. For the purposes of evaluating the maximum error, one may bear in mind $\mathcal{X} \simeq p$ and thus one may expand the cosine term into second order using a Taylor series. I assume the simple case of $b_{P,T} = 0$ so that $i_P = \pi/2$. I make further small-angle approximations to simplify the resultant expression for the error, which is justified since $2\pi t_I \ll P_P$. The other adjustment one needs to account for is that I have approximated $b_{P,T} = 0$ and $e_P = 0$. To generalize the result, I consider that the effect of $b_{P,T} > 0$ and $e_P > 0$ is to stretch or shrink the ingress/egress duration by a factor τ/τ_0. Therefore the expressions here are actually for $\mathcal{I}_0$, which may be written as $\mathcal{I}_0 = \mathcal{I}(\tau_0/\tau)$. One may now rewrite (5.36) as:

$$\sigma_{\check{F},\mathrm{obs}} > \left| \frac{\psi^{5/2}}{108m^3}\left[3\left(\sqrt{24mp - 9\psi} - 4\sqrt{2mp - \psi} - 6\sqrt{4mp - \psi} + \sqrt{8mp - \psi}\right)\right]\right| \tag{5.39}$$

where I have used:

$$\psi = \frac{2\pi (a_P/R_*)}{P_P}\frac{\kappa_0}{\kappa}\mathcal{I} \tag{5.40}$$

$$\frac{\kappa_0}{\kappa} \simeq \frac{\sqrt{1-b_{P,T}^2}\sqrt{1-e_{P^2}}}{\varrho_{P,T}} \tag{5.41}$$

Due to the approximations made, I find that this equation is only stable for $m \geq 2$. For any given data set, one simply needs to solve Equation (5.39) for m with some sensible estimates of $p, b_{P,T}, e_P, \omega_P, (a_P/R_*)$ and P_P. As an example, for Kepler-5b, taking the quoted parameters from the Koch et al. [46] paper, I find that even using $m = 2$ provides an error of 0.1 ppm, which is well below the typical measurement uncertainty of 130 ppm.

Limb-Darkened Trough-Curvature

Another part of the light curve where there is significant curvature, and thus one expects the maximum numerical integration error, is the limb-darkened light curve trough. However, the peak-to-peak size of the changes in flux induced by the limb darkening are much lower than the transit signal itself (i.e. δ); typically an order-of-magnitude. Further, the time scale over which these changes act is greater than that of the ingress/egress curvatures (i.e. $T_{14} \gg T_{12}$) except for grazing transits. So one can see that, in general, the errors in these numerical integration techniques will be dominated by the ingress/egress curvatures rather than the limb-darkening-induced light curve trough curvatures.

Contact Point Discontinuities

The final source of variation in the light curve gradient is that of the discontinuous change located at the contact points. Estimating the error due to this discontinuity may be achieved by assuming a trapezoid approximated light curve and considering the location where maximal error is induced. The largest error (and in fact only error) will occur for measurements close to contact points, or more specifically $|t_i - t_x| < \mathcal{I}/2$ where t_x is the time of one of the contact points.

Before the first contact point, the flux is a flat line at F=1 and after this point there is a linear slope with a gradient $-(\delta/T_{12})$. The error in Simpson's composite rule will depend upon the relative phasing between the centre of the integration and the contact point, i.e. $(t_i - t_I)$. Generalized to any phase, the true integrated flux of the trapezoid approximated light curve for the ith time stamp is given by:

$$\check{F}_{\mathrm{true},i} = 1 - \frac{\delta}{T_{12}}\frac{(2t_i + \mathcal{I})^2}{8\mathcal{I}} \tag{5.42}$$

For each value of $m = 1, 2, 3, \ldots$ I choose to set the phase to be such that the difference between the true integrated flux and that from Simpson's method is maximized. Under such a condition, it may be shown that the maximum error is given by:

$$\sigma_{\check{F}\,\text{Comp.Simp.}} = \frac{\mathcal{I}\delta}{24m^2 T_{12}} \tag{5.43}$$

$$= \frac{\mathcal{I}\delta}{6(N-1)^2 T_{12}} \tag{5.44}$$

For the system parameters of Kepler-5b, I find that using $m = 1, 2, 3$ induces a maximal error of 371, 93 and 41 ppm respectively. Given that the measurement uncertainties are 130 ppm [46], a suitable choice for the resolution would be $m=2$ since this means the maximum possible error of a data point in the least-favourable phasing would be below the measurement error.

It is interesting to see that for $m = 2$ the error was 0.1 ppm for the ingress/egress curvature of the same system, suggesting the discontinuity error dominates the error budget. Actually, this is expected from the arguments made earlier in this section. Therefore, in most applications, a selection for m based on the error induced by the contact point discontinuities will provide a robust integration resolution.

5.3.3.4 Resampling

An additional method for numerically integrating the light curve is discussed here. Consider a set of observations with integrated time stamps given by the vector $\check{\mathbf{t}}$. A second way of calculating $\check{\mathbf{F}}(\check{\mathbf{t}})$ is to resample the time vector into a very fine cadence, at which point one may assume $\check{\mathbf{F}} \simeq \mathbf{F}$. I define the temporary resampled time vector as $\check{\mathbf{t}}'$. As an example, for the *Kepler* data, one may choose to resample the 30 min integrations into 1 min integrations by expanding each time stamp, $\check{t}_i$ into a sub-vector of 30 equally spaced time stamps with a mean value given by $\check{t}_i$. The new temporary time array is used to generate a light curve using the normal Mandel and Agol [1] expressions giving $\mathbf{F}'(\check{\mathbf{t}}')$ (note that F here has no check sign because the Mandel and Agol [1] equations can only generate instantaneous flux, not integrated flux). The next step is to rebin the model light curve back to the original cadence to give $\mathbf{F}(\check{\mathbf{t}})$. Finally, I make the assumption $\check{\mathbf{F}}(\check{\mathbf{t}}') \simeq \mathbf{F}(\mathbf{t})$, i.e. the high cadence resampled time vector yields a light curve model consistent with a time vector of infinite cadence.

$$\check{\mathbf{t}} \underset{\text{resample}}{\rightarrow} \check{\mathbf{t}}' \underset{\text{MA02}}{\rightarrow} \mathbf{F}'(\check{\mathbf{t}}') \underset{\text{rebin}}{\rightarrow} \mathbf{F}(\check{\mathbf{t}}) \simeq \check{\mathbf{F}}(\check{\mathbf{t}}) \tag{5.45}$$

It can be seen that resampling into N sub-time stamps will increase the computation time by a factor of $\sim N$, since typically the Mandel and Agol [1] subroutine uses the majority of a light curve fitting algorithm's resources (especially for non-linear limb darkening). In the next subsubsection, I will show that the computation times can be decreased by "selective resampling".

One advantage of resampling is that one can choose to resample in such a way as to account for read-out and dead-times, which may be important if the instrument's duty cycle is quite poor.[9] The resampling of the ith time stamp into N sub-time stamps with labels $j = 1, 2, \ldots N-1, N$ can be expressed as:

$$t'_{i,j} = t_i + \left(j - \frac{N+1}{2}\right)\frac{\mathcal{I}}{N} \tag{5.46}$$

The flux of the ith time stamp is found by rebinning all N flux stamps from $j = 1$ to $j = N$.

$$\check{F}_i = \frac{\sum_{j=1}^{N} F'_{i,j}}{N} \tag{5.47}$$

Thus for the first few values of $N = 2$, $N = 3$ and $N = 4$ one would have:

$$\check{F}_i(N=2) = \frac{1}{2}\left[F(t_i - \mathcal{I}/4) + F(t_i + \mathcal{I}/4)\right] \tag{5.48}$$

$$\check{F}_i(N=3) = \frac{1}{3}\left[F(t_i - \mathcal{I}/3) + F(t_i) + F(t_i + \mathcal{I}/3)\right] \tag{5.49}$$

$$\check{F}_i(N=4) = \frac{1}{4}\left[F(t_i - 3\mathcal{I}/8) + F(t_i - \mathcal{I}/8) + F(t_i + \mathcal{I}/8) + F(t_i + 3\mathcal{I}/8)\right] \tag{5.50}$$

For a trapezoid approximated light curve, it can be easily shown that the error in these expressions, as a function of N, is given by:

$$\underset{\text{Resampling}}{\sigma_{\check{F}}} = \frac{\delta}{T_{12}} \frac{\mathcal{I}}{8N^2} \tag{5.51}$$

Therefore, the resampling method yields greater accuracy than the composite Simpson's method. In Kipping and Bakos [21], both the resampling and Simpson's composite rule were employed in completely independent analyses and the obtained results were consistent. Therefore, whilst one is free to use either method discussed here, the most efficient approach out of the two is resampling.

Gilliland et al. [48] reported that they used a method for fitting the long-cadence light curve of TreS-2b which I interpret to be equivalent to the resampling method. The authors split the LC intervals into 30 contributing sub-intervals corresponding to the SC cadence i.e. $N = 30$. For the reported LC RMS noise of 66 ppm and the system parameters of TreS-2b taken from Kipping and Bakos [3], I estimate that using $N = 5$ would produce a maximum possible error in the most unfavourably phased data point of 59 ppm and thus using $N = 30$ is excessive for this light curve. These equations therefore permit for a reduction in computational time of 600%. Such

[9] Note that this is not the case for *Kepler* which has a duty cycle of 91.4%.

a saving is highly advantageous in MCMC fitting, which is inherently expensive on the CPU.

5.3.3.5 Selective Resampling

Resampling time stamps which satisfy $|t_i - \tau_T| > (T_{14} + \mathcal{I})/2$ and $|t_i - \tau_O| > (O_{14} + \mathcal{I})/2$ is unnecessary since $\check{F}_i = F_i = 1$ in such cases (assuming one has folded multiple transits about the orbital period). I label this method of optimization as "selective resampling".

Since $(T_{14} + O_{14})/P_P \sim 2/(\pi(a_P/R_*))$, this can reduce the number of time stamps which require resampling by an order of magnitude for continuous staring telescopes like *Kepler* and *CoRoT*. It should be noted that selective resampling will not be possible if the light curve model includes phase variations of the planet e.g. HAT-P-7b, Borucki et al. [30].

5.3.4 Conclusions

I have explored how long-cadence data, with particular focus on *Kepler*, invalidates assumption A10 and consequently causes severe systematic errors in the retrieved physical parameters, unless accounted for. The effect is valid for any finite exposure time but increases with longer cadences. Long-cadence data smears out the light curve morphology, which acts to stretch out the ingress/egress duration and suppress limb darkening in the light curve trough. These two effects act to increase and decrease the retrieved impact parameter respectively. Critically, overestimating the impact parameter is shown to lead to severe underestimations of the stellar density which could lead to planetary candidates being rejected on the basis of being unphysical, in a similar manner as to what was found for eccentric orbits in Sect. 4.3.5.

Numerical integration techniques permit improved modelling of the transit light curve. I have discussed two particular methods, the composite Simpson's method and resampling. Expressions for estimating the errors of these techniques have been provided and I find that both methods produce an error which scales as N^{-2}, where N is the numerical resolution of the techniques. Out of these two discussed methods, the resampling approach yields a greater efficiency.

With corrective procedures for the invalidation of assumptions A0, A4/A5 and A10 now established, it is now possible to produce highly accurate transit light curve models. With the discussion and nuances of modelling the transit light curve of a single exoplanet now complete, it is possible to progress to the original aim of this thesis—-*methods to detect extrasolar moons in transiting systems*, which will be explored in Chap. 6.

References

1. K. Mandel, E. Agol, Analytic light curves for planetary transit searches. Astrophys. J. Lett. **580**, L171–L175 (2002)
2. D.M. Kipping, G. Tinetti, Nightside pollution of exoplanet transit depths. Mon. Not. R. Astron. Soc. **407**, 2589–2598 (2010)
3. D.M. Kipping, Binning is sinning: morphological light-curve distortions due to finite integration time. Mon. Not. R. Astron. Soc. **408**, 1758–1769 (2010)
4. D. Charbonneau, T.M. Brown, D.W. Latham, M. Mayor, Detection of planetary transits across a sun-like star. Astrophys. J. Lett. **529**, L45–L48 (2000)
5. G.W. Henry, G.W. Marcy, R.P. Butler, S.S. Vogt, A transiting "51 peg-like" planet. Astrophys. J. Lett. **529**, L41–L44 (2000)
6. T.M. Brown, D. Charbonneau, R.L. Gilliland, R.W. Noyes, A. Burrows, Hubble Space Telescope time-series photometry of the transiting planet of HD 209458. Astrophys. J. **552**, 699–709 (2001)
7. G. Bakos, R.W. Noyes, G. Kovács, K.Z. Stanek, D.D. Sasselov, I. Domsa, Wide-field millimagnitude photometry with the HAT: a tool for extrasolar planet detection. Publ. Astron. Soc. Pac. **116**, 266–277 (2004)
8. D.L. Pollacco, I. Skillen, A.C. Cameron, D.J. Christian, C. Hellier, J. Irwin, T.A. Lister, R.A. Street, R.G. West, D. Anderson, W.I. Clarkson, H. Deeg, B. Enoch, A. Evans, A. Fitzsimmons, C.A. Haswell, S. Hodgkin, K. Horne, S.R. Kane, F.P. Keenan, P.F.L. Maxted, A.J. Norton, J. Osborne, N.R. Parley, R.S.I. Ryans, B. Smalley, P.J. Wheatley, D.M. Wilson, The WASP project and the SuperWASP cameras. Publ. Astron. Soc. Pac. **118**, 1407–1418 (2006)
9. S. Seager, G. Mallén-Ornelas, A unique solution of planet and star parameters from an extrasolar planet transit light curve. Astrophys. J. **585**, 1038–1055 (2003)
10. D. Deming, S. Seager, L.J. Richardson, J. Harrington, Infrared radiation from an extrasolar planet. Nature **434**, 740–743 (2005)
11. J. Harrington, B.M. Hansen, S.H. Luszcz, S. Seager, D. Deming, K. Menou, J.Y.-K. Cho, L.J. Richardson, The phase-dependent infrared brightness of the extrasolar planet Upsilon Andromedae b. Science **314**, 623–626 (2006)
12. C.J. Grillmair, D. Charbonneau, A. Burrows, L. Armus, J. Stauffer, V. Meadows, Van J. Cleve, D. Levine, A spitzer spectrum of the exoplanet HD 189733b. Astrophys. J. Lett. **658**, L115–L118 (2007)
13. G. Tinetti, A. Vidal-Madjar, M.-C. Liang, J.-P. Beaulieu, Y. Yung, S. Carey, R.J. Barber, J. Tennyson, I. Ribas, N. Allard, G.E. Ballester, D.K. Sing, F. Selsis, Water vapour in the atmosphere of a transiting extrasolar planet. Nature **448**, 169–171 (2007)
14. M.R. Swain, G. Vasisht, G. Tinetti, The presence of methane in the atmosphere of an extrasolar planet. Nature **452**, 329–331 (2008)
15. M.R. Swain, G. Tinetti, G. Vasisht, P. Deroo, C. Griffith, J. Bouwman, P. Chen, Y. Yung, A. Burrows, L.R. Brown, J. Matthews, J.F. Rowe, R. Kuschnig, D. Angerhausen, Water, methane, and carbon dioxide present in the dayside spectrum of the exoplanet HD 209458b. Astrophys. J. **704**, 1616–1621 (2009)
16. S. Seager, D.D. Sasselov, Theoretical transmission spectra during extrasolar giant planet transits. Astrophys. J. **537**, 916–921 (2000)
17. G. Tinetti, J.-P. Beaulieu. The extrasolar planet atmosphere and exosphere: emission and transmission spectroscopy, in *IAU symposium*, vol. 253, pp. 231–237, February 2009. doi:10.1017/S1743921308026446
18. S. Seager, E.L. Turner, J. Schafer, E.B. Ford, Vegetation's red edge: a possible spectroscopic biosignature of extraterrestrial plants. Astrobiology **5**, 372–390 (2005)
19. D.W. Latham, W.J. Borucki, D.G. Koch, T.M. Brown, L.A. Buchhave, G. Basri, N.M. Batalha, D.A. Caldwell, W.D. Cochran, E.W. Dunham, G. Fűrész, T.N. Gautier, J.C. Geary, R.L. Gilliland, S.B. Howell, J.M. Jenkins, J.J. Lissauer, G.W. Marcy, D.G. Monet, J.F. Rowe,

D.D. Sasselov, Kepler-7b: a transiting planet with unusually low density. Astrophys. J. Lett. **713**, L140–L144 (2010)
20. H.A. Knutson, D. Charbonneau, L.E. Allen, J.J. Fortney, E. Agol, N.B. Cowan, A.P. Showman, C.S. Cooper, S.T. Megeath, A map of the day–night contrast of the extrasolar planet HD 189733b. Nature **447**, 183–186 (2007)
21. D.M. Kipping, G.Á. Bakos, Analysis of Kepler's short-cadence photometry for TrES-2b. ArXiv e-prints, June 2010
22. W.F. Welsh, J.A. Orosz, S. Seager, J.J. Fortney, J. Jenkins, J.F. Rowe, D. Koch, W.J. Borucki, The discovery of ellipsoidal variations in the Kepler light curve of HAT-P-7. Astrophys. J. Lett. **713**, 145–149 (2010)
23. E. Agol, N.B. Cowan, J. Bushong, H. Knutson, D. Charbonneau, D. Deming, J.H. Steffen, Transits and secondary eclipses of HD 189733 with Spitzer, in *IAU Symposium*, vol. 253, 2009, pp. 209–215
24. F. Pont, R.L. Gilliland, C. Moutou, D. Charbonneau, F. Bouchy, T.M. Brown, M. Mayor, D. Queloz, N. Santos, S. Udry, Hubble Space Telescope time-series photometry of the planetary transit of HD 189733: no moon, no rings, starspots. Astron. Astrophys. **476**, 1347–1355 (2007)
25. J.A. Dittmann, L.M. Close, E.M. Green, M. Fenwick, A tentative detection of a starspot during consecutive transits of an extrasolar planet from the ground: no evidence of a double transiting planet system around TrES-1. Astrophys. J. **701**, 756–763 (2009)
26. S. Czesla, K.F. Huber, U. Wolter, S. Schröter, J.H.M.M. Schmitt, How stellar activity affects the size estimates of extrasolar planets. Astron. Astrophys. **505**, 1277–1282 (2009)
27. E. Rauscher, K. Menou, J.Y.-K. Cho, S. Seager, B.M.S. Hansen, Hot Jupiter variability in eclipse depth. Astrophys. J. Lett. **662**, 115–118 (2007)
28. J.A. Orosz, P.H. Hauschildt, The use of the NextGen model atmospheres for cool giants in a light curve synthesis code. Astron. Astrophys. **364**, 265–281 (2000)
29. S.V. Jeffers, J.P. Aufdenberg, G.A.J. Hussain, A.C. Cameron, V.R. Holzwarth, Hubble Space Telescope observations of SV Cam—II. First derivative light-curve modelling using PHOENIX and ATLAS model atmospheres. Mon. Not. R. Astron. Soc. **367**, 1308–1316 (2006)
30. W.J. Borucki, D. Koch, J. Jenkins, D. Sasselov, R. Gilliland, N. Batalha, D.W. Latham, D. Caldwell, G. Basri, T. Brown, J. Christensen-Dalsgaard, W.D. Cochran, De E. Vore, E. Dunham, A.K. Dupree, T. Gautier, J. Geary, A. Gould, S. Howell, H. Kjeldsen, J. Lissauer, G. Marcy, S. Meibom, D. Morrison, J. Tarter, Kepler's optical phase curve of the exoplanet HAT-P-7b. Science **325**, 709 (2009)
31. D. Charbonneau, H.A. Knutson, T. Barman, L.E. Allen, M. Mayor, S.T. Megeath, D. Queloz, S. Udry, The broadband infrared emission spectrum of the exoplanet HD 189733b. Astrophys. J. **686**, 1341–1348 (2008)
32. E. Pfahl, P. Arras, B. Paxton, Ellipsoidal oscillations induced by substellar companions: a prospect for the Kepler mission. Astrophys. J. **679**, 783–796 (2008)
33. A. Claret, A new non-linear limb-darkening law for LTE stellar atmosphere models. Calculations for $-5.0 \leq \log[M/H] \leq +1$, $2000\,K \leq \text{Teff} \leq 50000\,\text{K}$ at several surface gravities. Astron. Astrophys. **363**, 1081–1190 (2000)
34. J.P. Beaulieu, D.M. Kipping, V. Batista, G. Tinetti, I. Ribas, S. Carey, J.A. Noriega-Crespo, C.A. Griffith, G. Campanella, S. Dong, J. Tennyson, R.J. Barber, P. Deroo, S.J. Fossey, D. Liang, M.R. Swain, Y. Yung, N. Allard, Water in HD 209458b's atmosphere from 3.6–8 microns IRAC photometric observations in primary transit. ArXiv e-prints September 2009.
35. F. Gauss, Zeitschrift fr Astronomie und verwandt Wissenschaften **1**, 187 (1817)
36. H.A. Knutson, D. Charbonneau, N.B. Cowan, J.J. Fortney, A.P. Showman, E. Agol, G.W. Henry, M.E. Everett, L.E. Allen, Multiwavelength constraints on the day–night circulation patterns of HD 189733b. Astrophys. J. **690**, 822–836 (2009)
37. T.M. Brown, Transmission spectra as diagnostics of extrasolar giant planet atmospheres. Astrophys. J. **553**, 1006–1026 (2001)
38. R.J. Barber, J. Tennyson, G.J. Harris, R.N. Tolchenov, A high-accuracy computed water line list. Mon. Not. R. Astron. Soc. **368**, 1087–1094 (2006)

39. L.S. Rothman, R.B. Wattson, R. Gamache, J.W. Schroeder, A. McCann, HITRAN HAWKS and HITEMP: high-temperature molecular database, in *Presented at the Society of Photo-Optical Instrumentation Engineers (SPIE) Conference Series*, vol. 2471, ed. by J.C. Dainty, 1995, pp. 105–111
40. S.A. Tashkun, V.I. Perevalov, J.-L. Teffo, A.D. Bykov, N.N. Lavrentieva, CDSD-1000, the high-temperature carbon dioxide spectroscopic databank. J. Quant. Spectrosc. Radiat. Transf. **82**, 165–196 (2003)
41. A. Borysow, U.G. Jorgensen, Y. Fu, High-temperature (1000–7000 K) collision-induced absorption of H"2 pairs computed from the first principles, with application to cool and dense stellar atmospheres. J. Quant. Spectrosc. Radiat. Transf. **68**, 235–255 (2001)
42. J.J. Fortney, M. Shabram, A.P. Showman, Y. Lian, R.S. Freedman, M.S. Marley, N.K. Lewis, Transmission spectra of three-dimensional hot Jupiter model atmospheres. Astrophys. J. **709**, 1396–1406 (2010)
43. J.P. Beaulieu, S. Carey, I. Ribas, G. Tinetti, Primary transit of the planet HD 189733b at 3.6 and 5.8 μm. Astrophys. J. **677**, 1343–1347 (2008)
44. J.-M. Désert, A. Lecavelierdes Etangs, G. Hébrard, D.K. Sing, D. Ehrenreich, R. Ferlet, A. Vidal-Madjar, Search for carbon monoxide in the atmosphere of the transiting exoplanet HD 189733b. Astrophys. J. **699**, 478–485 (2009)
45. G. Torres, J.N. Winn, M.J. Holman, Improved parameters for extrasolar transiting planets. Astrophys. J. **677**, 1324–1342 (2008)
46. D.G. Koch, W.J. Borucki, J.F. Rowe, N.M. Batalha, T.M. Brown, D.A. Caldwell, J. Caldwell, W.D. Cochran, E. DeVore, E.W. Dunham, A.K. Dupree, T.N. Gautier, J.C. Geary, R.L. Gilliland, S.B. Howell, J.M. Jenkins, D.W. Latham, J.J. Lissauer, G.W. Marcy, D. Morrison, J. Tarter, Discovery of the transiting planet Kepler-5b. Astrophys. J. Lett. **713**, L131–L135 (2010)
47. J.A. Carter, J.C. Yee, J. Eastman, B.S. Gaudi, J.N. Winn, Analytic approximations for transit light-curve observables, uncertainties, and covariances. Astrophys. J. **689**, 499–512 (2008)
48. R.L. Gilliland, J.M. Jenkins, W.J. Borucki, S.T. Bryson, D.A. Caldwell, B.D. Clarke, J.L. Dotson, M.R. Haas, J. Hall, T. Klaus, D. Koch, S. McCauliff, E.V. Quintana, J.D. Twicken, van J.E. Cleve, Initial characteristics of Kepler short cadence data. Astrophys. J. Lett. **713**, L160–L163 (2010). doi:10.1088/2041-8205/713/2/L160

Chapter 6
Transit Timing Effects Due to an Exomoon

The moving moon went up to the sky,
And nowhere did abide;
Softly she was going up,
And a star or two beside.
Samuel Taylor Coleridge, The Ancient Mariner (pt. IV)

6.1 Introduction

The methods and intricacies of modelling the light curve of a single transiting planet have now been covered. This necessary understanding enables the investigation of more complex situations involving three-bodies, rather than two. This therefore generalizes the model for cases where assumption A1 is invalid (see Sect. 3.3.1).

As discussed in Chap. 2, the moons of extrasolar planets present a unique and exciting challenge in modern astronomy. The motivations for such a search have been clearly established (Sect. 2.1) and with the transit now well understood (see Chaps. 3, 4 and 5), I begin here the discussion of what forms the ultimate goal of this thesis (Sect. 1.6)—a method to detect an exomoon in a transiting system.

6.2 Background Theory of Timing Deviations

6.2.1 The Sartoretti and Schneider Method

The first serious scientific investigation into a method for detecting exomoons in transiting systems comes from Sartoretti and Schneider [1]. This groundbreaking paper, published in the supplementary series of A&A, did not receive wide attention until very recently, where it has gone from an average of $\sim$4 citations per year from 2000 to 2009 up to 16 citations in 2010 (Fig. 6.1).

D. M. Kipping, *The Transits of Extrasolar Planets with Moons*, Springer Theses,
DOI: 10.1007/978-3-642-22269-6_6,

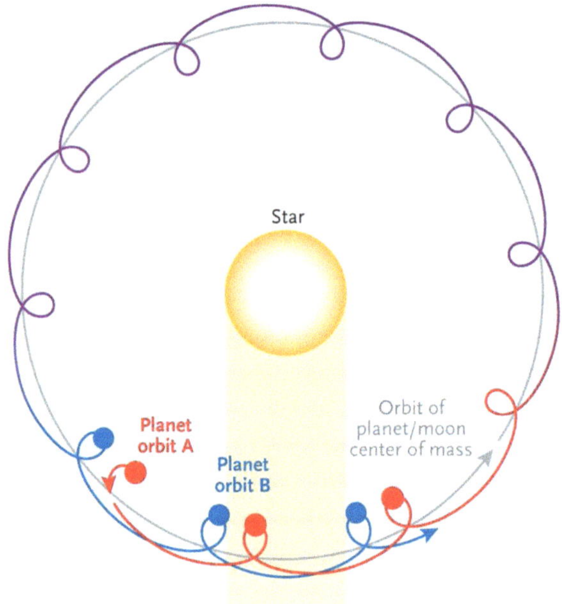

Fig. 6.1 Cartoon illustrating the composite motion of a planet in orbit of a star with an unseen exomoon companion. Unless the period of the moon happens to be resonant to the planet's orbital period (which there is no reason to expect) then orbit B will exhibit a different shift in position than orbit A at the time of inferior conjunction. These changes in position give rise to the TTV effect

Perhaps inspired by the astrometric method of detecting exoplanets, which in 1999 was considered to have more promise than it has since delivered, Sartoretti and Schneider [1] argued that even if an exomoon is too small to detect in transit, the gravitational perturbation which it induces on the host planet may not be. The very simple premise was that the planet and moon orbited a common centre-of-mass which itself orbited the star causing the planet to exhibit composite motion. This composite motion was discernible from a simple Keplerian orbit by virtue of the fact the position of the planet was offset from the barycentre and thus the times of transit minimum would also be displaced. Although not labelled as such in the paper, the phenomenon of changes in the times of transit minimum would soon be dubbed "transit time variation" (TTV).

6.2.2 Exomoon Transit Features

Before I provide Sartoretti and Schneider's mathematical formulation of TTV, I first mention that the authors also considered detecting moons through their transit features. These transit features could be due to either the moon eclipsing the planet (causing a flux increase) or the star (causing a flux decrease). This method is certainly highly viable for moons of large radius but modelling the light curves of planet–moon systems in an efficient manner is a very challenging problem.

To illustrate this complexity, consider that for a planet–star system it is possible to describe the light curve using just one parameter, S_{P*}, which can be in three possible states (see Sect. 3.3.1): (i) in-transit (ii) on-the-limb (iii) out-of-transit. Therefore, there exists 3 distinct cases, each with its own unique, analytic solution. In contrast, a planet–moon–star system requires three distances to characterize the light curve S_{P*}, S_{PS} and S_{S*} (where the "S" subscript denotes satellite). Since each one can be in three states again, there exists $3^3 = 27$ distinct cases, each with its own unique, analytic solution. Things are further exacerbated when one considers the effects of limb darkening and the three-body problem, which has no exact closed-form solution. As a result, modelling these light curves in an analytic manner is highly challenging and no proposed method currently exists in the literature (one could model the light curve numerically by pixelating the star, but such methods use inordinate amounts of CPU and memory making a fitting routine impractical).

As will be discussed in Chap. 8, part of the future work in exomoon detection theory will revolve around meeting this challenge. To this end, I have been developing a new analytic framework and computer code called `LUNA`, which accomplishes the goal of complete analytic modelling of the planet–moon–star transit light curve. However, this is currently a work in progress and does not constitute the main focus of this thesis. Therefore, for the remaining derivations I assume $(R_S/R_*)^2 \ll 1$ i.e. the moon does not induce a detectable transit signal. I label this assumption as $\alpha 1$. I also point out here that assumption $\alpha 0$ will be to consider the system comprised of a star with a single planet with a single moon.

6.2.3 The TTV of Sartoretti and Schneider

The authors explicitly assumed a coplanar configuration and implicitly assumed that the orbit of both the planet and satellite are circular, plus assumption $\alpha 1$.[1] There are other implicit assumptions which will be discussed in Sect. 6.3.4, and it is unclear whether the authors were aware of these or not. Sartoretti and Schneider [1] argued that the positional displacement of the planet away from the planet–moon barycentre is given by:

$$a_{PB} = (M_S/M_P)a_{SB} \tag{6.1}$$

where a_{PB} denotes the semi-major axis of the planet ("P") around the planet–moon barycentre ("B") and a_{SB} denotes the semi-major axis of the satellite ("S") around the same barycentre. The authors then argue that the velocity of the planet at the time of transit minimum is given by $v_{P,T} = 2\pi a_P/P_P$. Therefore, the maximum time shift of the planet away from that expected from a strictly linear ephemeris would be given by $(a_{PB}/v_{P,T})$ and the maximum possible observed difference between consecutive transits would be twice this value:

$$\Delta t \sim 2a_{SB}M_S M_P^{-1} \times P_P(2\pi a_P)^{-1} \tag{6.2}$$

[1] This last assumption is required or the moon's transit features would impair the timing of the planetary transit signal.

Equation 6.2 is given in the exact same format as that presented in the original paper, except for a change in notation to be consistent with that of this thesis. Equation 6.2 may be re-expressed in terms some typical astrophysical units and replacing $a_{SB} = \mathfrak{D}' R_{\text{Hill}}$, where R_{Hill} is the Hill radius, I find:

$$\begin{aligned} \Delta t &\sim \frac{\mathfrak{D}' M_S P_P}{3^{1/3} \pi M_P^{2/3} M_*^{1/3}} \\ &= 36.0\mathfrak{D}' \left(\frac{M_S}{M_\oplus}\right) \left(\frac{P_P}{\text{years}}\right) \left(\frac{M_J}{M_P}\right)^{2/3} \left(\frac{M_\odot}{M_*}\right)^{1/3} \text{min} \end{aligned} \tag{6.3}$$

This 36 min deviation illustrates the exciting potential of this technique to detect an exomoon. After this pioneering work, it seemed as though the methodology for detecting an exomoon was solved, but unfortunately problems with this TTV technique soon became apparent, as discussed next in Sect. 6.2.4.

6.2.4 The Problems with Detecting Exomoons Using TTV

6.2.4.1 Problem 1: A Multitude of Phenomena

As was mentioned earlier, Sartoretti and Schneider [1] devised the TTV method for detecting moons, although it was not called TTV at the time. Six years later, Agol et al. [2] and Holman and Murray [3] almost simultaneously predicted that if a system has two or more planets then TTV would be induced due to planet–planet gravitational interactions (this appears to be the first instance of referring to the effect as TTV). These effects are particularly pronounced for planets in mean motion resonance (MMR), where timing deviations of hours can be induced [4].

Unfortunately for exomoons, this new prediction of the phenomenon meant that if one observed TTVs then it could be due to either a moon or a perturbing planet. This was exacerbated later when a barrage of papers predicted a plethora of possible sources for TTV, including:

- Trojan bodies [5]
- Parallax effects [6]
- Kozai mechanism induced eccentricity variation [7]
- Apsidal precession [8]
- Star spots [9]
- Stellar proper motion [10]
- Planetary in-fall [11]
- The Applegate effect [12]
- Stellar binarity [13]

With so many effects potentially responsible for TTV, the detection of TTVs no longer seemed viable as an unambiguous detection tool for exomoons. This can be summarized by the desideratum for some kind of unique signature of the exomoon's timing deviations, which would distinguish it from the other effects.

6.2.4.2 Problem 2: Undersampling

The second major problem with the TTV due to an exomoon is undersampling. I begin this discussion by considering the Hill radius. R_{Hill} represents the maximum possible stable orbital distance of an exomoon away from its host planet. As discussed earlier in Sect. 2.2.3, the true limit is actually less than R_{Hill} [14]. Let us therefore denote the semi-major axis of the satellite around the planet as some fraction, $\mathfrak{D}$, of the Hill radius, in a similar way to as was done in (6.3):

$$a_{SP} = \mathfrak{D} a_P \left(\frac{M_P}{3M_*}\right)^{1/3} \tag{6.4}$$

Now I use Kepler's Third Law to replace the distances with orbital periods:

$$G^{1/3}(M_P + M_S)^{1/3}\left(\frac{P_S}{2\pi}\right)^{2/3} = \mathfrak{D} G^{1/3}(M_* + M_P + M_S)^{1/3}\left(\frac{P_P}{2\pi}\right)^{2/3}\left(\frac{M_P}{3M_*}\right)^{1/3} \tag{6.5}$$

Now I assume $M_* \gg M_P \gg M_S$, giving:

$$P_S = P_P \mathfrak{D}^{3/2}\left(\frac{1}{3}\right)^{1/2} \tag{6.6}$$

Equation 6.6 is very powerful, constraining a satellite's orbital period purely in terms of $\mathfrak{D}$ and P_P. I have been unable to find this expression in the previous literature and believe it may be a novel result, first pointed out in Kipping [15]. Since $\mathfrak{D} < 1$ for all bounded moons, then $P_S < 0.57735 P_P$.

The sampling rate of the exomoon TTV signal is once every transit i.e. P_P. In order to avoid aliasing, one is required to sample the signal at above the Nyquist rate i.e. $2/P_S = 3.46/P_P$. However, it is physically impossible to sample the signal 3.46 times per P_P, the best one can ever do is once[2] per P_P. Therefore, the observed TTVs caused by an exomoon *will always be undersampled*, in most cases heavily so. As a result, a Fourier periodogram of the TTVs would reveal a rich forest of possible harmonic frequencies with no way of distinguishing which signal is the correct one.

Despite not being able to retrieve the correct period, the deviations themselves will still be present and cause an excess variance in the residuals of the transit times from a linear ephemeris. This excess variance is detectable if the standard deviation of the scatter is much greater than the errors bars of the measurements themselves.

Since it is possible to measure the amplitude through excess scatter, then one can evaluate:

---

[2] Hypothetically, one could actually sample the signal twice per P_P using the occultation, but the error on this timing would be much higher. In any case, even sampling of twice per P_P would still cause aliasing.

$$M_S a_{SB} = \frac{\pi a_P M_P}{2 P_P} \sqrt{2} \delta_{\mathrm{TTV}} \tag{6.7}$$

where δ_{TTV} is the RMS amplitude of the TTV signal. Therefore, the final problem of undersampling becomes evident—one can only measure $M_S a_{SB}$ i.e. mass multiplied by distance, a.k.a. "the moment" of the exomoon.

6.3 An Updated Model for TTV

6.3.1 Introduction

Two major problems exist with the proposed TTV method of Sartoretti and Schneider [1] and a hypothetical wish list of required breakthroughs would read as: (1) the ability to distinguish between an exomoon's TTV and other sources, (2) the ability to measure both the exomoon mass and orbital distance. In the remainder of this chapter, I provide a new method of detecting exomoons which achieves both of these goals and thus makes transit timing a viable method for detecting exomoons and thus coming full circle to satisfy the original goal of this thesis. This remainder of this chapter is based upon the papers [15, 16], with some new updates.

Before I provide this new method, I will first provide an updated model for the transit timing variations (TTV) of an exomoon, including the effects of orbital eccentricity and inclination of both the planet and moon plus the mutual longitude of the ascending node. I will then go on to predict and introduce a new observable, transit duration variations (TDV) and compare it to TTV.

I will begin by outlining the assumptions of the model presented here. So far, only one assumption has been made, $\alpha 1$): the satellite induces no transit features $\Rightarrow (R_S/R_*)^2 \ll 1$. In order to proceed, the first requirement is a coordinate system for the positions of the planet and the moon which will require an additional approximation, given that no general analytic solution exists for the three-body problem.

6.3.2 The Nested Two-Body Model

Since there is a planet, star and moon all in play, one must deal with the three-body problem. The three-body problem is quite daunting given that no analytic, general solution has ever been found. In practice, one is forced to either use numerical methods or some restricted case through an approximation. In this work, I will adopt what I label as "the nested two-body" model. In this framework, one has the planet–moon forming a close-binary which orbits the star at a much greater distance. The basic premise is that in the frame of reference of the planet–moon barycentre, both bodies follow Keplerian orbits. Then, in the reference frame of the star, the motions can be considered as a composite of these local Keplerian orbits on top of the global Keplerian motion around the star.

The advantages of the nested two-body model are (a) it is simple to describe analytically since one only has Keplerian orbits, (b) more general than the restricted circular three-body problem, which insists eccentricities are zero, (c) particularly applicable for the problem of interest—motion of a planet over short timescales (i.e. we do not care about long term precessions, secular changes, etc). I will here derive the conditions under which the nested two-body model is a good approximation for the three-body problem.

For a planet–moon pair at infinite orbital separation, the validity of the nested model can be understood intuitionally. As the planet–moon pair moves in, the disturbance due to the star becomes greater and the approximation will break down. One may follow the standard treatment used in lunar theory [17] to calculate the disturbance to this approximation. I select a non-rotating reference frame with the star at rest at the origin, for which one may treat the frame as inertial since the mass of the star is much greater than the planet or moon. The satellite's equation of motion is given by:

$$\ddot{\mathbf{r}}_S = -n_S^2 a_S^3 \frac{(\mathbf{r}_S - \mathbf{r}_P)}{|\mathbf{r}_S - \mathbf{r}_P|^3} - n_P^2 a_P^3 \frac{\mathbf{r}_S}{|\mathbf{r}_S|^3} \tag{6.8}$$

where $\mathbf{r}_S$ and $\mathbf{r}_P$ are the position vectors of the satellite and planet relative to the star, n_P is the mean-motion of the planet around the star and a_P is the corresponding semi-major axis. Similarly, n_S is the mean-motion of the moon around the planet and a_S is the corresponding semi-major axis. To translate to the reference frame of the planet, one may substitute:

$$\mathbf{r} = \mathbf{r}_S - \mathbf{r}_P \tag{6.9}$$

$$\mathbf{r}' = -\mathbf{r}_P \tag{6.10}$$

These vectors now describe the positions of the satellite and the star, respectively, relative to the planet. It follows that in this non-inertial frame where the planet is at rest, but the coordinate axes point in fixed directions, the equation of motion for the satellite becomes:

$$\ddot{\mathbf{r}} = -n_S^2 a_S^3 \frac{\mathbf{r}}{|\mathbf{r}|^3} + n_P^2 a_P^3 \left[\frac{(\mathbf{r}' - \mathbf{r})}{|\mathbf{r}' - \mathbf{r}|^3} - \frac{\mathbf{r}'}{|\mathbf{r}'|^3} \right] \tag{6.11}$$

The term to the right of the addition sign represents the disturbance to Kepler's Third Law from a simple nested two-body approximation. The disturbing function can be written as:

$$\begin{aligned}
\phi &= n_P^2 a_P^3 \left[\frac{\mathbf{r}_S}{|\mathbf{r}_S|^3} - \frac{\mathbf{r}_P}{|\mathbf{r}_P|^3} \right] \\
|\phi| &\simeq n_P^2 a_P \big([1 + (a_S/a_P)]^{-2} - 1 \big) \\
|\phi| &\simeq 2 n_P^2 a_S
\end{aligned} \tag{6.12}$$

where I have assumed a low eccentricity system and expanded to first order for $(a_S/a_P) \ll 1$. For the disturbing function to be small, one therefore requires:

$$\begin{aligned} \left| n_S^2 a_S^3 \frac{\mathbf{r}}{|\mathbf{r}|^3} \right| &\gg |\phi| \\ n_S^2 &\gg 2n_P^2 \end{aligned} \tag{6.13}$$

where the final line gives the condition under which this derivation is ultimately valid and is equivalent to:

$$P_S^2 \ll P_P^2/2 \tag{6.14}$$

Using (6.6), $(P_S/P_P) \simeq \sqrt{\mathfrak{D}^3/3}$, this constraint becomes assumption $\alpha 2$:

$$\mathfrak{D}^3 \ll 3/2 \tag{6.15}$$

Defining $\ll$ to indicate an order-of-magnitude difference, i.e. a factor of 10, this constrains $\mathfrak{D} \leq 0.531$. Note that this distance is larger than that predicted as the maximum stable separation for a prograde satellite of $\mathfrak{D} = 0.4895$ [14]. However, a retrograde moon can be stable at up to $\mathfrak{D} = 0.9309$, at which point $\mathfrak{D}^3$ is only about one half of 3/2 and thus Keplerian motion is not guaranteed.

6.3.3 Coordinate System

In the nested two body problem, the planet–moon barycentre orbits the star in a Keplerian orbit and thus one can see that the planet–moon barycentre, denoted by the subscript "B", replaces the planet ("P") in (3.6). The coordinates of the barycentre are therefore given by:

$$\begin{aligned} X_B &= r_P \cos(\omega_P + f_P) \\ Y_B &= r_P \sin(\omega_P + f_P) \cos i_P \\ Z_B &= r_P \sin(\omega_P + f_P) \sin i_P \end{aligned} \tag{6.16}$$

Furthermore, the satellite orbits the barycentre of the planet–moon system with a semi-major axis of a_{SB} in a Keplerian orbit. The motion is therefore highly analogous to the case derived earlier for a planet–star system. Indeed, in the reference frame of the planet–moon barycentre, one may simply adapt (3.6), replacing the subscripts for the planet to that of the satellite:

$$\begin{aligned} X_{SB} &= r_{SB}[\cos\Omega_S \cos(\omega_S + f_S) - \sin i_S \sin\Omega_S \sin(\omega_P + f_S)] \\ Y_{SB} &= r_{SB}[\sin\Omega_S \cos(\omega_S + f_S) + \sin i_S \cos\Omega_S \sin(\omega_P + f_S)] \\ Z_{SB} &= r_{SB} \cos i_S \sin(\omega_S + f_S) \end{aligned} \tag{6.17}$$

where $r_{SB} = a_{SB}(1 - e_S^2)/(1 + e_P \cos f_S)$. The planet's reflex motion is simply given by:

$$\begin{aligned}
X_{PB} &= -(M_S/M_P)X_{SB} = -r_{PB}[\cos\Omega_S \cos(\omega_S + f_S) - \sin i_S \sin\Omega_S \sin(\omega_P + f_S)] \\
Y_{PB} &= -(M_S/M_P)Y_{SB} = -r_{PB}[\sin\Omega_S \cos(\omega_S + f_S) + \sin i_S \cos\Omega_S \sin(\omega_P + f_S)] \\
Z_{PB} &= -(M_S/M_P)Z_{SB} = -r_{PB} \cos i_S \sin(\omega_S + f_S)
\end{aligned} \tag{6.18}$$

where $r_{PB} = (M_S/M_P)r_{SB}$. These coordinates define the motion of the planet in the reference frame of the planet–moon barycentre. However, the reference frame of interest is that of the star at the origin. Proceeding through the rotations for ω_P and i_P (Ω_P rotation is not performed as the transit is insensitive to this parameter) as was done in Sect. 3.3.1, the final sky-projected coordinates of the planet are given by:

$$\begin{aligned}
X_P = r_P \cos(f_P + \omega_P) + r_{PB}\,[&-\cos(f_S + \omega_S)\cos(\omega_P + \Omega_S) \\
&+ \sin i_S \sin(f_S + \omega_S)\sin(\omega_P + \Omega_S)]
\end{aligned} \tag{6.19}$$

$$\begin{aligned}
Y_P = r_P \sin(f_P + \omega_P)\cos i_P + r_{PB}\,[&\sin(f_S + \omega_S)[\cos i_S \sin i_P \\
&- \cos i_P \sin i_S \cos(\omega_P + \Omega_S)] - \cos i_P \sin(\omega_P + \Omega_S)\cos(f_S + \omega_S)]
\end{aligned} \tag{6.20}$$

$$\begin{aligned}
Z_P = r_P \sin(f_P + \omega_P)\sin i_P - r_{PB}\,[&\sin(f_S + \omega_S)[\cos i_S \cos i_P \\
&+ \sin i_P \sin i_S \cos(\omega_P + \Omega_S)] + \sin i_P \sin(\omega_P + \Omega_S)\cos(f_S + \omega_S)]
\end{aligned} \tag{6.21}$$

There are two things to note from these equations. First of all, $X_{PB} \neq (X_P - X_B)$, and equivalently for Y_{PB} and Z_{PB}. I will therefore define $\Delta X_P = (X_P - X_B)$, $\Delta Y_P = (Y_P - Y_B)$ and $\Delta Z_P = (Z_P - Z_B)$. Second, all instances of Ω_S occur in the form $(\omega_P + \Omega_S)$. I will define this quantity as $\varpi_S = \omega_P + \Omega_S$, the "longitude of the periapsis", in the rest of this work.

6.3.4 Model Assumptions

6.3.4.1 The Importance of Declaring the Model Assumptions

Curiously, the TTV of an exomoon is very easy to write down with some simple intuitive feel but actually characterizing what the assumptions are and under what conditions this toy model is valid is more subtle. In this subsection, I will provide a strict outline of the model and the required assumptions, no matter how seemingly obvious. This is extremely important as it has already been seen how breaking model assumptions for the transit model causes significant departures (Sect. 3.3.7). Further, identifying the assumptions allows one to calculate the range of moons and planets for which it is actually legitimate to conduct TTV analyses on.

6.3.4.2 RMS Amplitude Definition

The first thing to establish is that whatever form is derived for the TTV effect, it is only ever possible to infer the RMS amplitude, as discussed in Sect. 6.2.4. Therefore, regardless as to the functional form, the following integral must be performed to calculate the RMS amplitude:

$$\delta_{\mathrm{TTV}} = \sqrt{\frac{1}{2\pi} \int_{f_S=0}^{2\pi} [\mathrm{TTV}(f_S)]^2 \, \mathrm{d}f_S} \tag{6.22}$$

I outline this point early on, because naturally if the TTV has an elaborate functional form, the integral will become increasingly challenging to compute analytically, and perhaps not even possible. Therefore, it is required to seek a relatively simple but nevertheless physically accurate and justified form for the TTV.

6.3.4.3 Two Definitions of the TTV

One subtlety with TTV is that there actually exists two definitions of what constitutes the TTV. The first definition could be considered as the pure theoretical definition whilst the second is the observable or measurable TTV. I will discuss the first definition initially, which will make it clear why a second definition exists.

Theoretical Definition of the TTV. In the nested two-body problem, the planet–moon barycentre orbits the star in a Keplerian orbit. Therefore the instant when the barycentre satisfies $\mathrm{d}S_{B*}/\mathrm{d}t = 0$ (i.e. the equivalent of the transit minimum) occurs at approximately the time of inferior conjunction, as before. These events are therefore separated in time by exactly P_P and have a strictly linear ephemeris.

In contrast, the instant when the transit minimum of the planet occurs will be given by when $\mathrm{d}S_{P*}/\mathrm{d}t = 0$, which will be a function of f_S. In other words, the transit minima do not occur at the same instances as the barycentric minima. Therefore, the transit minima do not follow a strictly linear ephemeris. The temporal deviation of the transit minima away from the barycentric minima is the source of the TTV (for both definitions).

The theoretical definition for the TTV of a transit event would therefore be to calculate the difference between the times $t(\dot{S}_{B*} = 0)$ and $t(\dot{S}_{P*} = 0)$:

$$\mathrm{TTV} = t(\dot{S}_{P*} = 0) - t(\dot{S}_{B*} = 0) \tag{6.23}$$

Observable Definition of the TTV. The theoretical definition classes the TTV as the time difference between the instants of the barycentric and planetary transit minima. However, consider the case of a moon moving on a very short orbital period, say less than the transit duration. Under such a circumstance, the planet–moon separation would be very different between the start and the end of the transit. When one fits such a light curve, the timing deviation would be essentially the averaged-out temporal deviation across the whole duration. In other words, the motion of the moon smears out the TTV.

It can be seen that a derivation of the theoretical TTV is required before it is possible to account for the practicalities of smearing effects. I therefore proceed here to consider what model assumptions are required for the theoretical TTV. The observational TTV will be discussed in Sect. 6.3.8.

6.3.4.4 Assumptions $\alpha3$ and $\alpha4$: Slow-Moving Moon

The theoretical definition of the TTV is based upon solving the instants when the sky-projected separations hit minima. However, it has been shown earlier (see Sect. 4.1) that solving the exact time when $\dot{S}_{B*} = 0$ is non-trivial and possesses no simple analytic form. The mathematics will be even more complex for the case of $\dot{S}_{P*} = 0$. Therefore, one can easily appreciate that approximations will be necessary to make this calculation possible.

The first two approximations I therefore make are $\alpha3$) the sky-projected distance between the barycentre and the planet does not change appreciably over the timescale of the TTV amplitude $\alpha4$) the sky-projected velocity of the planet does not change appreciably over the timescale of the TTV amplitude. Under such approximations, the TTV definition becomes:

$$\mathrm{TTV} = \frac{[S_{P*} - S_{B*}](t(f_P = f_{P,T}))}{\frac{\mathrm{d}[S_{P*}]}{\mathrm{d}t}(t(f_P = f_{P,T}))} \tag{6.24}$$

where the parentheses after each term mean that the function is evaluated at the time, t, when $f_P = f_{P,T}$ i.e. the transit minimum. The approximations $\alpha3$ and $\alpha4$ will be written out mathematically shortly, but first it is useful to adopt additional approximations. Firstly, when trying to solve the true anomaly at the instant of the transit minimum for the simple case of a moonless planet and a star, it was seen that a simple, closed-form solution is elusive (Sect. 4.1). In light of this, one can easily appreciate that solving for $f_{P,T}$ with the moon included will be even worse. Therefore, a useful approximation to make is that the numerator and denominator in (6.24) experience negligible change when evaluated at $f_P = (\pi/2 - \omega_P)$ rather than $f_{P,T}$.

Since I have assumed the distance is constant over the timescale of the TTV amplitude, assumption $\alpha3$, then assuming $[S_{P*} - S_{B*}](t(f_P = f_{P,T})) \simeq [S_{P*} - S_{B*}](t(f_P = \pi/2 - \omega_P))$ is actually already implicit by assumption $\alpha3$ and no further assumption is needed. This is because the timescale of the TTV amplitude is approximately equal to that for the planet to move between $f_P = f_{P,T}$ and $f_P = (\pi/2 - \omega_P)$.

The same is also true for the denominator, if one wishes to assume the velocity at $f_P = f_{P,T}$ and $f_P = (\pi/2 - \omega_P)$ are approximately equal, then this is equivalent to assuming that the velocity does not change appreciably over the timescale of the TTV effect i.e. assumption $\alpha4$. This line of argument leads to the following definition for the TTV effect:

$$\mathrm{TTV} = \frac{[S_{P*} - S_{B*}](t(f_P = \pi/2 - \omega_P))}{\frac{\mathrm{d}[S_{P*}]}{\mathrm{d}t}(t(\pi/2 - \omega_P))} \tag{6.25}$$

6.3.4.5 Assumptions $\alpha 5$ and $\alpha 6$: Motion is in the $\hat{X}$-Direction

Now, it has been established that the key term of interest is the sky-projected distance between the planet and the planet–moon barycentre i.e. $(S_{P*} - S_{B*})$. I will write out this distance in terms of the Cartesian elements:

$$
\begin{aligned}
S_{P*} - S_{B*} &= \frac{\sqrt{X_P^2 + Y_P^2}}{R_*} - \frac{\sqrt{X_B^2 + Y_B^2}}{R_*} \\
&= \frac{\sqrt{(X_B + \Delta X_P)^2 + (Y_P + \Delta Y_P)^2}}{R_*} - \frac{\sqrt{X_B^2 + Y_B^2}}{R_*}
\end{aligned} \tag{6.26}
$$

The presence of terms with "B" subscripts is non-preferable as one can intuitionally understand that it is the distance between the barycentre and the planet (i.e. the "Δ" terms) which ultimately dominate the TTV. Inspecting the above, it is clear that if one assumes the motion is predominantly in the $\hat{X}$-direction, then the Y-terms can be ignored and the following simplification would be possible:

$$
\begin{aligned}
S_{P*} - S_{B*} &\simeq \frac{\sqrt{(X_B + \Delta X_P)^2}}{R_*} - \frac{\sqrt{X_B^2}}{R_*} \\
S_{P*} - S_{B*} &\simeq \frac{\Delta X_P}{R_*} = \frac{X_P - X_B}{R_*}
\end{aligned} \tag{6.27}
$$

These simplifications can be seen to greatly aid in the analysis and thus are adopted in this work and denoted as assumptions $\alpha 5$ and $\alpha 6$. The TTV equation now becomes:

$$
\mathrm{TTV} = \frac{[X_P - X_B](t(f_P = \pi/2 - \omega_P))}{\frac{\mathrm{d}[S_{P*}]}{\mathrm{d}t}(t(\pi/2 - \omega_P))} \tag{6.28}
$$

The mathematical form of assumption $\alpha 5$ yields:

$$
\begin{aligned}
&|Y_B| \ll |X_B| \\
\Rightarrow &|\cos i_P| \ll 1
\end{aligned} \tag{6.29}
$$

Assumption $\alpha 6$ can be seen to be equivalent to this assumption if one assumes $r_{PB} \ll r_P$, which gives:

$$
\begin{aligned}
&a_{PB} \ll a_P \\
&\mathfrak{D} \ll \frac{3^{1/3} M_*^{1/3} M_P^{2/3}}{M_S} \\
&\mathfrak{D} \ll 4655.19 \left(\frac{M_*}{M_\odot}\right)^{1/3} \left(\frac{M_P}{M_J}\right)^{2/3} \left(\frac{M_\oplus}{M_S}\right)
\end{aligned} \tag{6.30}
$$

The above assumption can be seen to be essentially valid in virtually all conditions. Therefore since $a_{PB} \ll a_P$ in almost all cases and $\cos i_P \ll 1$ by definition for a transit to occur, then assumptions $\alpha 5$ and $\alpha 6$ are excellent approximations.

6.3.4.6 Assumption $\alpha 7$: Barycentric Dominated Velocity

The next step is to compute the velocity-like term. The $\dot{X}_P(t(f_P = \pi/2 - \omega_P))$ term contains an f_S dependency and this causes problems later when it is necessary to perform the integration over f_S, as required by (6.22). I have not been able to solve said integral if the velocity-like term has an f_S dependency. Therefore, a required approximation is that $\dot{X}_P \simeq \dot{X}_B$. This is equivalent to assuming that $|\dot{X}_B(t(f_P = \pi/2 - \omega_P))| \gg |\Delta\dot{X}_P|(t(f_P = \pi/2 - \omega_P))$ i.e. the barycentric velocity is much greater than the planet's reflex motion, at the instant of inferior conjunction. This forms assumption $\alpha 7$. Differentiating, and setting $f_P = (\pi/2 - \omega_P)$, I find:

$$\begin{aligned}
&|\dot{X}_B(t(f_P = \pi/2 - \omega_P))| \gg |\Delta\dot{X}_P(t(f_P = \pi/2 - \omega_P))| \\
&\quad \frac{a_P(1 + e_P \sin\omega_P)}{\sqrt{1 - e_P^2} P_P} \gg \frac{a_{PB}}{\sqrt{1 - e_S^2} P_S} \\
&\quad \times |\cos\varpi_S[e_S \sin\omega_S + \sin(f_S + \omega_S)] + \sin i_S \sin\varpi_S[e_S \cos\omega_S + \cos(f_S + \omega_S)]|
\end{aligned} \tag{6.31}$$

The terms inside the large modulus contain all of the f_S dependency. In order to evaluate the limits under which the above inequality is valid, one is interested in setting the terms inside the large modulus to be the maximum possible value. I differentiated the expression with respect to f_S and then solved for when the result is equal to zero. This yields four solutions for $f_{S,\max}$. I then generated 10^6 Monte Carlo simulations where all of the parameters are randomly varied and four versions of the term in the modulus are calculated each time, for each solution for $f_{S,\max}$. I then compute the maximum of each of these four lists of possible maxima which gives $\{1.99424, 1.98929, 1.99485, 1.99275\}$. Therefore, it is reasonable to conclude that the maximum possible value is equal to 2. Accordingly, the assumption $\alpha 7$ becomes:

$$\frac{a_P(1 + e_P \sin\omega_P)}{\sqrt{1 - e_P^2} P_P} \gg \frac{2a_{PB}}{\sqrt{1 - e_S^2} P_S} \tag{6.32}$$

Now using Kepler's Third Law and 6.6, one obtains:

$$\mathfrak{D}^{1/2} \gg 2 \times 3^{1/6} \left(\frac{M_S}{M_*^{1/3} M_P^{2/3}} \right) \left(\frac{\sqrt{1 - e_P^2}}{\sqrt{1 - e_S^2}(1 + e_P \sin\omega_P)} \right) \tag{6.33}$$

For near circular orbits, and assuming $\gg$ to be an order-of-magnitude, this becomes:

$$\mathfrak{D} \geq 0.000055 \left(\frac{M_S}{M_\oplus} \right)^2 \left(\frac{M_\odot}{M_*} \right)^{2/3} \left(\frac{M_J}{M_P} \right)^{4/3} \tag{6.34}$$

So this constraint will be satisfied in essentially all cases as well.

6.3.5 Valid Range for the Theory TTV Model

I have now established an upper constraint on $\mathfrak{D}$ using assumption $\alpha 2$ and a lower limit from $\alpha 7$. Therefore there is a permitted bounded range of $\mathfrak{D}$ which can be investigated. Before I do so, I will calculate the conditions which assumptions $\alpha 3$ and $\alpha 4$ imply. For $\alpha 3$, the condition is that $(S_{P*} - S_{B*})$ does not change appreciably over the timescale of the TTV amplitude. Under assumption $\alpha 5$ this becomes that $(X_P - X_B) = \Delta X_P$ does not change appreciably over the same time. ΔX_P is characterized by the distance a_{PB} and the TTV amplitude is approximately given by the Sartoretti and Schneider [1] model, given in (6.2):

$$\frac{\mathrm{d}\Delta X_P}{\mathrm{d}t} \ll a_{PB}\left[2a_{SB}M_S M_P^{-1} \times P_P(2\pi a_P)^{-1}\right]^{-1}$$
$$\mathfrak{D}^{1/2} \gg \frac{4\times 3^{1/6}}{\sqrt{1-e_S^2}}\frac{M_S}{M_P^{2/3}M_*1/3} \tag{6.35}$$
$$\mathfrak{D}^{1/2} \gg 0.00149\left(\frac{M_S}{M_\oplus}\right)\left(\frac{M_J}{M_P}\right)^{2/3}\left(\frac{M_\odot}{M_*}\right)^{1/3}$$

where on the last line I have assumed $e_S \ll 1$. If one defines $\gg$ to indicate an order of magnitude then the above becomes:

$$\mathfrak{D} \geq 0.00012\left(\frac{M_S}{M_\oplus}\right)^2\left(\frac{M_J}{M_P}\right)^{4/3}\left(\frac{M_\odot}{M_*}\right)^{2/3} \tag{6.36}$$

Therefore, just like $\alpha 7$, assumptions $\alpha 3$ places a lower limit on $\mathfrak{D}$ for TTV studies. The two have the same functional dependency but $\alpha 3$ places a more stringent lower constraint on $\mathfrak{D}$ (by a factor of 2) and so represents the more realistic bounding limit.

For $\alpha 4$, the constraint bounds $\dot{X}_B$ to exhibit negligible variation over the timescale of the TTV amplitude.

$$\left.\frac{\mathrm{d}^2 X_B}{\mathrm{d}t^2}\right|_{f_P=\pi-2\omega_P} \delta_{\mathrm{TTV}} \ll \left.\frac{\mathrm{d}X_B}{\mathrm{d}t}\right|_{f_P=\pi-2\omega_P} \tag{6.37}$$

However, since $\ddot{X}_B(f_P = \pi/2 - \omega_P) = 0$, then the above can be seen to be valid in all cases. It has now been established that $\alpha 7$ places a lower limit and $\alpha 2$ places an upper limit on $\mathfrak{D}$, which together imply:

$$0.00012\left(\frac{M_S}{M_\oplus}\right)^2\left(\frac{M_J}{M_P}\right)^{4/3}\left(\frac{M_\odot}{M_*}\right)^{2/3} \leq \mathfrak{D} \leq \left(\frac{3}{20}\right)^{1/3} \tag{6.38}$$

It can therefore be seen that the theoretical TTV model presented in this thesis is invalid if the LHS exceeds the RHS of this inequality. This condition gives rise to the following condition, which must be satisfied for a TTV search using the model of this thesis to be possible:

$$M_S < (48.98 M_\oplus) \left(\frac{M_P}{M_J}\right)^{1/3} \left(\frac{M_\odot}{M_*}\right)^{1/3} \frac{1}{\sqrt{1-e_S^2}} \tag{6.39}$$

Therefore, the model for the theory TTV presented here is generally valid for all prograde satellites with terrestrial masses. The final list of assumptions which are used, and their respective consequential conditions which must be satisfied, are (conditions given for low-eccentricity systems):

- $\alpha 0$) There is only one moon and one planet in the system
- $\alpha 1$) The moon does not impart any transit features onto the light curve $\Rightarrow R_S^2 \ll R_*^2$
- $\alpha 2$) The three-body problem may be described with the nested two-body model $\Rightarrow \mathfrak{D}^3 \ll 3/2$
- $\alpha 3$) The change in the sky-projected planet–moon separation is negligible during the transit event $\Rightarrow \mathfrak{D}^{1/2} \gg 4 \times 3^{1/6} M_S M_*^{-1/3} M_P^{-2/3}$
- $\alpha 4$) The change in the sky-projected velocity of the planet is negligible during the transit event $\Rightarrow$ no further constraints
- $\alpha 5$) The sky-projected planet moves predominantly in the $\hat{X}$-direction $\Rightarrow a_{PB} \ll a_P$
- $\alpha 6$) The sky-projected planet–moon barycentre moves predominantly in the $\hat{X}$-direction $\Rightarrow |\cos i_P| \ll 1$
- $\alpha 7$) The sky-projected reflex motion of the planet is much less than the sky-projected barycentric motion $\Rightarrow \mathfrak{D}^{1/2} \gg 2 \times 3^{1/6} M_S M_*^{-1/3} M_P^{-2/3}$

6.3.6 TTV Waveform and Amplitude

Under assumptions $\alpha 0 \rightarrow \alpha 7$, the temporal deviation of the planet away from the planet–moon barycentre's position at the time of inferior conjunction is given by:

$$\text{TTV} = \frac{[X_P - X_B](t(f_P = \pi/2 - \omega_P))}{\frac{\mathrm{d}[X_B]}{\mathrm{d}t}(t(\pi/2 - \omega_P))} \tag{6.40}$$

Now one requires evaluating $[X_P - X_B](t(f_P = \pi/2 - \omega_P))$:

$$[X_P - X_B](t(f_P = \pi/2 - \omega_P)) = r_{PB} [-\cos(f_S + \omega_S)\cos\varpi_S + \sin i_S \sin(f_S + \omega_S)\sin\varpi_S] \tag{6.41}$$

And the sky-projected velocity of the planet–moon barycentre, in the $\hat{X}$-direction, is given by:

$$
\begin{aligned}
\frac{\mathrm{d}[X_B]}{\mathrm{d}t}(t(\pi/2-\omega_P)) &= \frac{\mathrm{d}X_B}{\mathrm{d}f_P}\bigg|_{f_P=\pi/2-\omega_P} \frac{\mathrm{d}f_P}{\mathrm{d}t}\bigg|_{f_P=\pi/2-\omega_P} \\
&= \Bigg[a_P \frac{1-e_P^2}{(1+e_P\cos f_P)^2}[e_P\sin\omega_P + \sin(f_P+\omega_P)] \\
&\quad \times \frac{2\pi}{P_P}\frac{(1+e_P\cos f_P)^2}{(1-e_P^2)^{3/2}}\Bigg]_{f_P=\pi/2-\omega_P} \\
&= \frac{2\pi a_P}{P_P}\frac{1+e_P\sin\omega_P}{(1-e_P^2)^{1/2}}
\end{aligned}
\tag{6.42}
$$

The final TTV equation is therefore:

$$
\mathrm{TTV}(f_S) \simeq \left[\frac{a_S\sqrt{1-e_P^2}(1-e_S^2)M_S P_P}{2\pi a_P M_P(1+e_P\sin\omega_P)}\right]\Lambda_{\mathrm{TTV}}(f_S) \tag{6.43}
$$

$$
\Lambda_{\mathrm{TTV}}(f_S) = \left[\frac{\cos\varpi_S\cos(f_S+\omega_S) - \sin i_S\sin\varpi_S\sin(f_S+\omega_S)}{1+e_S\cos f_S}\right] \tag{6.44}
$$

The TTV effect is therefore a function of the true anomaly of the satellite, as expected. In general, the satellite will take a different true anomaly every time one measures a transit. Therefore, one can obtain the RMS amplitude of the TTV effect using (6.22). Integrating this equation, I obtain:

$$
\delta_{\mathrm{TTV}} = \frac{a_S M_S P_P}{a_P M_P}\frac{(1-e_S^2)\sqrt{1-e_P^2}}{(1+e_P\sin\omega_P)}\sqrt{\frac{\Phi_{\mathrm{TTV}}}{2\pi}} \tag{6.45}
$$

$$
\begin{aligned}
\Phi_{\mathrm{TTV}} = \frac{2\pi}{(1-e_S^2)^2\left[1+\sqrt{1-e_S^2}\right]} &\Bigg[\cos^2\omega_S\Bigg(\sin^2 i_S\sin^2\varpi_S(1-e_S^2)^{3/2} \\
&\qquad + \cos^2\varpi_S\left(1+e_S^2\left[\sqrt{1-e_S^2}-1\right]\right)\Bigg) \\
+ \sin^2\omega_S\Bigg(\sin^2 i_S\sin^2\varpi_S&\left(1+e_S^2\left[\sqrt{1-e_S^2}-1\right]\right) \\
&\qquad + (1-e_S^2)^{3/2}\cos^2\varpi_S\Bigg) \\
+\frac{1}{2}\sin i_S\sin 2\omega_S\sin 2\varpi_S&\left(e_S^2[1-2\sqrt{1-e_S^2}]+\sqrt{1-e_S^2}-1\right)\Bigg]
\end{aligned}
\tag{6.46}
$$

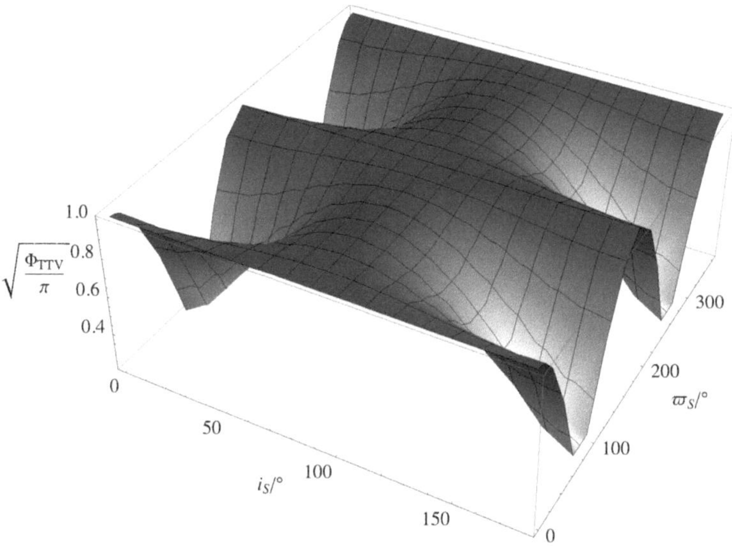

Fig. 6.2 Dependency of Φ_{TTV}, the TTV enhancement factor, on an exomoon's orbital inclination, i_S, and longitude of the periapsis, ϖ_S, as measured from the plane of the planetary orbit. Maximum TTVs occur for co-aligned systems. In this plot, I have assumed a circular orbit for both the exomoon and the host planet

6.3.7 Properties of the TTV RMS Amplitude

In the limit of circular and co-aligned orbits, as was considered by Sartoretti and Schneider [1], $\Phi_{\text{TTV}} \to \pi$ and thus the RMS amplitude agrees with that found by the original paper. However, the advantages of this much more general expression for the TTV waveform and RMS amplitude are manifold. In Fig. 6.2, I show the dependency of $\sqrt{\Phi_{\text{TTV}}/\pi}$, which is essentially the scaling factor for the TTV amplitude, in the case of a circular lunar orbit. In this case, a very simple expression for Φ_{TTV} is possible:

$$\lim_{e_S \to 0} \Phi_{\text{TTV}} = \pi \left(1 - \cos^2 i_S \sin^2 \varpi_S\right) \tag{6.47}$$

Equation 6.47 and Fig. 6.2 reveal that exomoons possessing $|i_S - \pi/2| > 0$ or $\varpi_S > 0$ only act to decrease the TTV amplitude and are thus less favourable for detection.

Note, that the equations presented here are also improvements upon the model presented in [15, 16], where longitude of the ascending node was fixed to zero and it was approximated that the velocity was equal to the tangential barycentric orbital velocity, rather than the sky-projected velocity of the barycentre in the $\hat{X}$-direction.

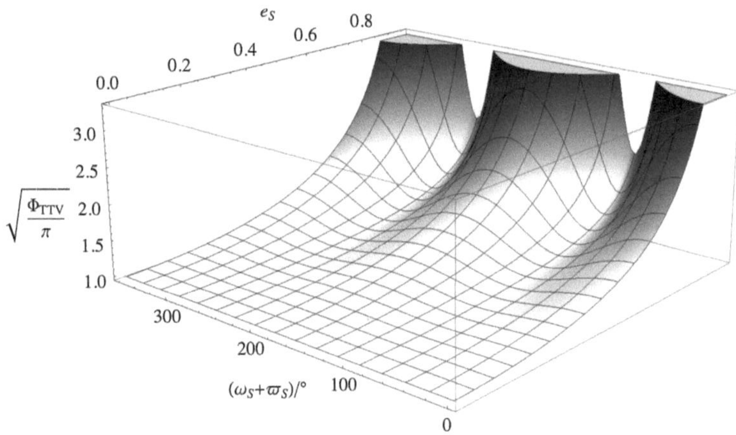

Fig. 6.3 Dependency of Φ_{TTV}, the TTV enhancement factor, on an exomoon's eccentricity and the sum of the argument of periapsis and the longitude of periapsis, $\omega_S + \varpi_S$. In this plot, I have assumed $i_S = \pi/2$

These improvements mean that the above equations should be used rather than those previously presented, for the greatest accuracy.

The effects of eccentricity, e_S, are illustrated in Fig. 6.3. Here, I assume $i_S = \pi/2$ which gives:

$$\lim_{i_S \to \pi/2} \Phi_{\mathrm{TTV}} = \pi \left[\frac{2\left(1 - e_S^2\right)\cos^2(\omega_S + \varpi_S) + \sqrt{1 - e_S^2}\left[1 + \left(-1 + 2e_S^2\right)\cos 2(\omega_S + \varpi_S)\right]}{(1 - e_S^2)^2\left[1 + \sqrt{1 - e_S^2}\right]} \right] \tag{6.48}$$

Equation 6.48 is plotted in Fig. 6.3 and reveals how eccentric moons can cause strong enhancements to the TTV amplitude. However, in general, one does not expect highly eccentric systems exomoons to be dynamically stable [14].

6.3.8 Observational TTV

The model presented thus far has been for the theoretical TTV, which ignores the practicalities of fitting light curves. As discussed in Sect. 6.3.4, a real determination of the timing deviation of the transit would differ from the theory definition if the moon was moving quickly enough. This motion would cause a blurring or smearing of the signal.

At the simplest level, this could be considered by extending assumptions $\alpha 3$ and $\alpha 4$ to hold over the timescale of the transit duration rather than the TTV amplitude. However, this would cause the equations to hold at a much higher accuracy than the TTV is ever likely to be measured to. One has to acknowledge that moons will induce very small observable effects and so the measurement of a TTV signal to, say, signal-to-noise SNR = 10 would have to be considered quite optimistic, in general. Therefore, if the smearing effect induces an error much less than 10%, then it would be safe to assume the observational TTV is equivalent to the theory TTV.

Out of the two assumptions, $\alpha 4$ is less constraining than $\alpha 3$. This is because the sky-projected acceleration of the barycentre is zero at the time of inferior conjunction and thus the assumption of constant velocity is excellent for both the timescale of the TTV amplitude and the timescale of the transit duration. I will denote the extension of $\alpha 4$ to the larger timescale as $\alpha 4*$, which as discussed here has the same conditions as $\alpha 4$. With this point established, I will turn the focus onto $\alpha 3$ and its extension to $\alpha 3*$.

In light of the acknowledgement that there exists a measurement error in real observations, I will consider the extension of $\alpha 3$, which I denote as $\alpha 3*$, to be that *the smearing effect of the lunar motion causes a change in the TTV amplitude which is much less than the measurement error*. If one had a measurement error of σ_{TTV}, then the acceptable tolerance level for the error in the TTV equation would be given by σ_{theory}:

$$\begin{aligned} &\sqrt{\sigma_{\mathrm{theory}}^2 + \sigma_{\mathrm{TTV}}^2} - \sigma_{\mathrm{TTV}} \ll \sigma_{\mathrm{TTV}} \\ &\Rightarrow \sigma_{\mathrm{theory}} \ll 2\sigma_{\mathrm{TTV}} \end{aligned} \tag{6.49}$$

So if for example $\sigma_{\mathrm{TTV}} = 10\%$, then an acceptable tolerance level in the smearing effect would be $\sigma_{\mathrm{theory}} \simeq 20\%$. The error in the theory TTV is controlled by the error in assuming $[X_P - X_B] = [X_P - X_B](t(f_P = \pi/2 - \omega_P))$. This may be evaluated by computing $\Delta X_P(t(f_P = \pi/2 - \omega_P - \Delta f_P/2)) - \Delta X_P(t(f_P = \pi/2 - \omega_P))$:

$$\sigma_{\mathrm{theory}} \simeq \frac{\Delta X_P(t(f_P = \pi/2 - \omega_P - \Delta f_P/2)) - \Delta X_P(t(f_P = \pi/2 - \omega_P))}{a_{PB}} \tag{6.50}$$

The two instances in time correspond to the ingress up to the moment of inferior conjunction and so are separated by $\tilde{T}/2$. If one assumes the satellite as a nearly circular orbit, then $\Delta \mathfrak{M}_S \simeq \Delta f_S$ and so $\Delta f_S \simeq \pi \tilde{T}/P_S$. The maximum smearing occurs when the velocity is largest which occurs when $\Delta X_P = 0$. This, in turn, occurs when $f_S + \omega_S = \tan^{-1}[1/\sin i_S \tan \varpi_S]$ for a low eccentricity orbit. If one assumes $\sin i_S \simeq 1$ (i.e. a coplanar moon) then this becomes $(f_S + \omega_S) = (\pi/2 - \varpi_S)$. It is now possible to evaluate σ_{theory}:

$$\sigma_{\mathrm{theory}} \simeq \sin\left[\frac{\pi \tilde{T}}{P_S}\right] \tag{6.51}$$

The requirement of $\alpha 3*$ then becomes:

$$\sin\left[\frac{\pi \tilde{T}}{P_S}\right] \ll 2\sigma_{\mathrm{TTV}} \tag{6.52}$$

Further expansion is possible by writing out $\tilde{T}$ assuming that $(a_P/R_*)^2 \varrho_{P,T}^2 \gg b_{P,T}^2$ and employing a small-angle approximation and finally using (6.6) plus that $\gg$ indicates an order-of-magnitude:

$$\mathfrak{D} \geq 0.196 \left(\frac{10\%}{\sigma_{\mathrm{TTV}}}\right)^{2/3} \left(\frac{10}{(a_P/R_*)}\right)^{2/3} (1 - b_{P,T}^2)^{1/3} \left(\frac{\sqrt{1-e_P^2}}{1+e_P \sin\omega_P}\right)^{2/3} \tag{6.53}$$

Equation 6.53 reveals that for typical transiters with $(a_P/R_*) > 10$, the constraint will be satisfied across the majority of the Hill sphere. For $(a_P/R_*) > 100$, which is the likely requirement for a moon to be dynamically stable, this translates to $\mathfrak{D} > 0.042$. Having established that the observational TTV is equal to the theory TTV under assumption $\alpha 3*$, the model for the TTV due to an exomoon is complete.

6.3.9 Valid Range for the Observational TTV

Just as was done in Sect. 6.3.5, the new lower bound from $\alpha 3*$ may be combined with the upper bound from $\alpha 2$ to estimate the feasible range for observational TTV (where I here assume near-circular orbits):

$$\frac{a_P}{R_*} > 0.224 \frac{\sqrt{1 - b_{P,T}^2}}{\sigma_{\mathrm{TTV}}} \tag{6.54}$$

6.4 Velocity-Induced Transit Duration Variation

6.4.1 The TDV-V Waveform and Amplitude

The theory of TTV is now complete, but as pointed out in Sect. 6.2.4, it is not sufficient as a method for the unambiguous detection of an exomoon. TTV is conceptually analogous to astrometry in that one is looking for a companion to a massive body by searching for the reflex motion and consequent variations in position. Astrometry is well-known to be highly complementary to the radial velocity technique and thus this line of thought brings one to consider whether variations in the velocity of a planet could betray the presence of a moon. Such variations would induce transit duration variations (TDV).

TDV has been previously been discussed as a possible test of general relativity by [8]. In this case, the changes were secular in nature whereas one would expect the reflex motion due to a moon to cause periodic variations of the same frequency as the TTV. In this section, I consider the TDV due to an exomoon and conclude that it should produce a detectable signal.

An exomoon's TDV may be found by considering that the duration of a transit is inversely proportional to the sky-projected velocity of the planet across the star. Since I am only considering TDV to be due to velocity variations, I label this TDV effect as TDV-V, where the "V" stands for velocity. The required assumptions are $\alpha 0$, $\alpha 1$, $\alpha 2$, $\alpha 4*$, $\alpha 5$, $\alpha 6$ and $\alpha 7$. The planetary transit duration scales as:

$$\tilde{T}_P \propto \left[\frac{\mathrm{d}X_P}{\mathrm{d}t} \bigg|_{f_P=\pi/2-\omega_P} \right]^{-1} \tag{6.55}$$

The presence of a moon means that $\mathrm{d}X_P/\mathrm{d}t \neq \mathrm{d}X_B/\mathrm{d}t$. The deviation of the duration away from the barycentric motion is given by:

$$\begin{aligned} \text{TDV-V} &= \tilde{T}_P - \tilde{T}_B \\ &= \tilde{T}_B \left[\frac{\tilde{T}_P}{\tilde{T}_B} - 1 \right] \\ &= \tilde{T}_B \left[\frac{\frac{\mathrm{d}X_B}{\mathrm{d}t}\Big|_{f_P=\pi/2-\omega_P}}{\frac{\mathrm{d}X_P}{\mathrm{d}t}\Big|_{f_P=\pi/2-\omega_P}} - 1 \right] \end{aligned} \tag{6.56}$$

The differentiated position of the planet–moon barycentre is given by:

$$\frac{\mathrm{d}X_B}{\mathrm{d}t}\bigg|_{f_P=\pi/2-\omega_P} = \left(\frac{2\pi a_P}{P_P} \right) \left(\frac{1 + e_P \sin\omega_P}{\sqrt{1-e_P^2}} \right) \tag{6.57}$$

From (6.21), one may write $X_P = X_B + \Delta X_P$ where ΔX_P is the perturbation term due to the moon and may be expressed independent of f_P. Therefore:

$$\frac{\mathrm{d}X_P}{\mathrm{d}t}\bigg|_{f_P=\pi/2-\omega_P} = \frac{\mathrm{d}X_B}{\mathrm{d}t}\bigg|_{f_P=\pi/2-\omega_P} + \frac{\mathrm{d}\Delta X_P}{\mathrm{d}t}\bigg|_{f_P=\pi/2-\omega_P} \tag{6.58}$$

where the first term on the RHS is already known. The second term is:

$$\begin{aligned} \frac{\mathrm{d}\Delta X_P}{\mathrm{d}t} &= \frac{\mathrm{d}\Delta X_P}{\mathrm{d}f_S}\frac{\mathrm{d}f_S}{\mathrm{d}t} \\ &= \frac{2\pi a_S}{\sqrt{1-e_S^2}P_S} [\cos\varpi_S\,[e_S \sin\omega_S + \sin(f_S+\omega_S)] \\ &\quad + \sin i_S \sin\varpi_S\,[e_S\cos\omega_S + \cos(f_S+\omega_S)]] \end{aligned} \tag{6.59}$$

Note, that since X_{PB} has no dependency on f_P, then evaluating the above at the instant of inferior conjunction does not actually change the equation at all. One may now bring assumption $\alpha 7$ into play:

$$\frac{\mathrm{d}\Delta X_P}{\mathrm{d}t} \ll \left.\frac{\mathrm{d}X_B}{\mathrm{d}t}\right|_{f_P=\pi/2-\omega_P} \tag{6.60}$$

With this approximation, a first-order expansion of the TDV-V waveform is given by:

$$\text{TDV-V} = -\frac{\frac{\mathrm{d}(X_P-X_B)}{\mathrm{d}t}}{\left.\frac{\mathrm{d}X_B}{\mathrm{d}t}\right|_{f_P=\pi/2-\omega_P}}\tilde{T}_B \tag{6.61}$$

$$= \tilde{T}_B\left(\frac{a_S M_S P_P}{a_P M_P P_S}\right)\left(\frac{\sqrt{1-e_P^2}}{\sqrt{1-e_S^2}(1+e_P\sin\omega_P)}\right)\Lambda_{\text{TDV-V}}(f_S) \tag{6.62}$$

$$\begin{aligned}\Lambda_{\text{TDV-V}}(f_S) &= \cos\varpi_S[e_S\sin\omega_S+\sin(f_S+\omega_S)]\\ &\quad+\sin i_S\sin\varpi_S[e_S\cos\omega_S+\cos(f_S+\omega_S)]\end{aligned} \tag{6.63}$$

Integrating over f_S, as was done for (6.46), one obtains the RMS TDV-V amplitude:

$$\begin{aligned}\delta_{\text{TDV-V}} &= \sqrt{\frac{1}{2\pi}\int_{f_S=0}^{2\pi}[\text{TDV-V}(f_S)]^2\,\mathrm{d}f_S}\\ \delta_{\text{TDV-V}} &= \tilde{T}_B\left(\frac{a_S M_S P_P}{a_P M_P P_S}\right)\left(\frac{\sqrt{1-e_P^2}}{\sqrt{1-e_S^2}(1+e_P\sin\omega_P)}\right)\sqrt{\frac{\Phi_{\text{TDV-V}}}{2\pi}}\end{aligned} \tag{6.64}$$

$$\begin{aligned}\Phi_{\text{TDV-V}} = \frac{\pi}{8}\Big[&-2e_S^2\cos 2\omega_S[1+3\cos 2\varpi_S]\\ &+(1+e_S^2)[6+\cos 2(i_S-\varpi_S)+2\cos 2\varpi_S+\cos 2(i_S+\varpi_S)]\\ &-2\cos 2i_S\left[1+e_S^2+2e_S^2\cos 2\omega_S\sin^2\varpi_S\right]+8e_S^2\sin i_S\sin 2\omega_S\sin 2\varpi_S\Big]\end{aligned} \tag{6.65}$$

As with Φ_{TTV}, $\Phi_{\text{TDV-V}} \to \pi$ for circular, co-aligned orbits. As was provided for the TTV effect, it is useful to write the TDV-V amplitude in astrophysical units:

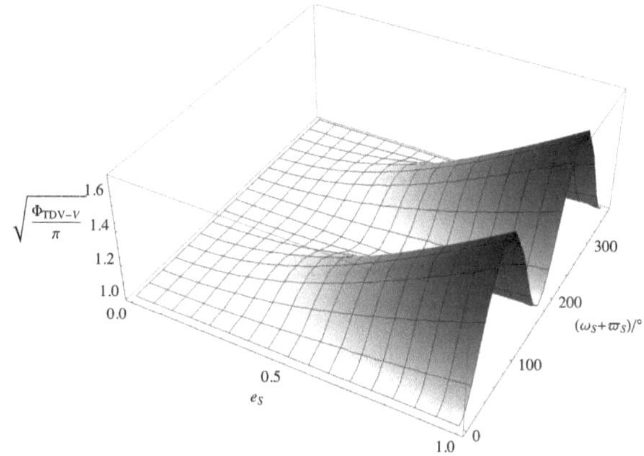

Fig. 6.4 Dependency of $\Phi_{\text{TDV-V}}$, the TDV-V enhancement factor, on an exomoon's eccentricity, e_S, and the sum of the argument of periapsis and the longitude of the periapsis, $(\omega_S + \varpi_S)$, In this plot, I have assumed $i_S = \pi/2$

$$\delta_{\text{TDV-V}} = (13.4\,\text{s}) \left(\frac{1}{\mathfrak{D}}\right)^{1/2} \left(\frac{\tilde{T}_B}{10\,\text{h}}\right) \left(\frac{M_S}{M_\oplus}\right) \left(\frac{M_J}{M_P}\right)^{2/3} \left(\frac{M_\odot}{M_*}\right)^{1/3} \times \left(\frac{\sqrt{1-e_P^2}}{\sqrt{1-e_S^2}(1+e_P \sin\omega_P)}\right) \sqrt{\frac{\Phi_{\text{TDV-V}}}{2\pi}} \tag{6.66}$$

6.4.2 Properties of the TDV-V RMS Amplitude

As was done in Sect. 6.3.7, I will here investigate the properties of the derived TDV-V RMS amplitude. In the case of a moon on a circular orbit, a very simple expression for $\Phi_{\text{TDV-V}}$ is possible, which happens to be equivalent to Φ_{TTV} for the same conditions.

$$\lim_{e_S \to 0} \Phi_{\text{TDV-V}} = \pi \left(1 - \cos^2 i_S \sin^2 \varpi_S\right) = \lim_{e_S \to 0} \Phi_{\text{TTV}} \tag{6.67}$$

Since this function is plotted in Fig. 6.2, it will not be repeated here. The effects of eccentricity, e_S are illustrated in Fig. 6.4. As was done for the TTV case, I set $i_S = \pi/2$ which simplifies $\Phi_{\text{TDV-V}}$ to:

$$\lim_{i_S \to \pi/2} \Phi_{\text{TDV-V}} = \pi \left(1 + 2e_S^2 \sin^2(\omega_S + \varpi_S)\right) \tag{6.68}$$

Figure 6.4 plots (6.68), which shows how eccentricity always leads to an enhancement of the TDV-V amplitude. As for the TTV case, the function is total controlled by one single angle, given by the sum of the argument of periapsis and the longitude of the periapsis.

6.5 TTV and TDV-V as Complementary Methods

6.5.1 *Mass and Period Determination*

I here define η as the ratio of the TDV to the TTV RMS amplitudes. Each of these amplitudes are derivable through searches for excess variance and so η is readily measurable.

$$\begin{aligned}\eta &= \frac{\delta_{\mathrm{TDV}}}{\delta_{\mathrm{TTV}}}\\ &= \frac{\tilde{T}_B}{P_S}\frac{1}{(1-e_S^2)^{3/2}}\sqrt{\frac{\Phi_{\mathrm{TDV\text{-}V}}}{\Phi_{\mathrm{TTV}}}}\end{aligned} \tag{6.69}$$

If one makes the reasonable approximation that $e_S \ll 1$, which is expected for dynamically stable exomoons [14], then one may make use of the fact:

$$\lim_{e_S \to 0} \Phi_{\mathrm{TTV}} = \lim_{e_S \to 0} \Phi_{\mathrm{TDV\text{-}V}} = \pi\left(1 - \cos^2 i_S \sin^2 \varpi_S\right) \tag{6.70}$$

Therefore, for $e_S \ll 1$, (6.69) becomes:

$$\lim_{e_S \to 0} \eta = \frac{\tilde{T}_B}{P_S} \tag{6.71}$$

The elegant (6.71) shows that the measurement of a TTV and TDV amplitude due to an exomoon allows one to directly determine P_S. Armed with P_S and Kepler's Third Law, one may use either δ_{TTV} or $\delta_{\mathrm{TDV\text{-}V}}$ to obtain M_S as well.

A further possibility to use the derived period to look for the closest harmonic frequency in the periodogram which would further constrain P_S and consequently M_S. Alternatively, one could use the harmonic period to estimate $\sqrt{(1-e_S^2)^{-3}\Phi_{\mathrm{TDV\text{-}V}}/\Phi_{\mathrm{TTV}}}$, which essentially characterizes how non-circular and non-coplanar the moon is.

Therefore, combining TTV and TDV together allows one to determine both the mass and period of the exomoon, which satisfies the first critical desideratum outlined in Sect. 6.2.4. Now just one problem remains—the requirement for a unique signature of an exomoon.

6.5.2 Phase Difference

Consider the simple case of $i_S \simeq \pi/2$ and $\Omega_S \simeq 0$ and $e_S = e_P = 0$. These conditions may seem limited but actually are dynamically expected as the Hill sphere of stable moon orbits shrink as the eccentricity and inclination angles move away from coplanarity [18]. Under this simple circumstance, the waveform components of the two signals become:

$$\begin{aligned} \lim_{e_P,e_S\to 0}\lim_{\Omega_S\to 0}\lim_{i_S\to\pi/2} \Lambda_{\text{TTV}} &= \cos f_S \\ \lim_{e_P,e_S\to 0}\lim_{\Omega_S\to 0}\lim_{i_S\to\pi/2} \Lambda_{\text{TDV-V}} &= \sin f_S \end{aligned} \tag{6.72}$$

Therefore, the TDV-V leads TTV by a $\pi/2$ phase shift. This phase difference is paramount—it is the key to unlocking an exomoon detection through timing effects. By detecting both signals and observing this phase shift, the detection would be unambiguous against all other phenomenon, since presently no other effect is predicted to induce such a phase shift. In fact, no other phenomena are predicted to even induce both periodic (and detectable) TTVs *and* TDVs.

The origin of the phase shift can be understood by considering that the planet's local Keplerian orbit appears as simple harmonic motion (SHM) on the sky-plane. Since TTV is a position effect and TDV-V is a velocity effect, then just like in SHM with a swinging pendulum, the velocity and position will always be 90° out of phase, and so too are the TTV and TDV-V effects.

In practice, the effects of orbital eccentricity and non-coalignment distort the phase difference away from $\pi/2$. However, one expects such systems to be relatively rare, due to the contracted regions of stability [18]. Therefore, the existence of this phase shift acts as a unique signature for an exomoon and discriminates the timing signals from other phenomena.

For a retrograde orbit, $i_S \simeq 3\pi/2$, but the phase shift is preserved. This tell us that the ratio of the TDV-V to the TTV effect provides no information about the sense of orbital motion of the moon:

$$\begin{aligned} \lim_{e_P,e_S\to 0}\lim_{\Omega_S\to 0}\lim_{i_S\to 3\pi/2} \Lambda_{\text{TTV}} &= \cos f_S \\ \lim_{e_P,e_S\to 0}\lim_{\Omega_S\to 0}\lim_{i_S\to 3\pi/2} \Lambda_{\text{TDV-V}} &= \sin f_S \end{aligned} \tag{6.73}$$

With the ability to determine both the exomoon mass and discriminate it against other phenomena, both of the problems described with TTV alone have been solved Sect. 6.2.4. TDV-V can therefore be seen to be the key to solving the problem of detecting exomoons through timing effects. With this, I have now met the original goal of this thesis—a method to detect exomoons in transiting systems. However, in the next section, I will discuss a additional complication which arises for non-coplanar configurations.

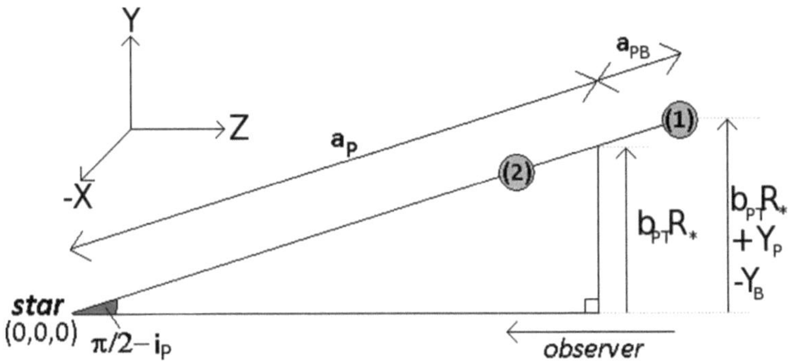

Fig. 6.5 Cartoon of the TDV-TIP effect. Here, the moon is relaxed into the same orbital plane as the planet's orbit and causes the planet to experience reflex motion illustrated by the two positions of the planet, (1) and (2). This motion can be seen to cause a change in the apparent impact parameter, which causes a change in the transit duration

6.6 Transit Impact Parameter Induced Transit Duration Variations

6.6.1 Definition

Not long after the prediction of TDV-V in [15], a follow-on paper soon predicted that an additional component to TDV should also exist for non-coplanar systems [16]. The premise was that if the plane of the planet–moon orbit was inclined to the line-of-sight, then the planet would have a non-zero component of motion in the $\hat{Y}$-direction i.e. $|\Delta Y_P| > 0$ (illustrated in Fig. 6.5). This motion would change the transit impact parameter and since the transit duration is highly sensitive to this term, then another form of TDV should occur.

Transit impact parameter induced transit duration variations (TDV-TIP) can be completely understood in terms of ΔY_P and how it modifies the apparent transit impact parameter. The modified transit duration, using the $\tilde{T}^{\mathrm{one}}$ approximation, would be given by:

$$\tilde{T}_P^{\mathrm{TIP}} = \frac{P_P}{\pi} \frac{\varrho_{P,T}^2}{\sqrt{1-e_P^2}} \arcsin \sqrt{\frac{1-[b_{p,T}+(\Delta Y_P/R_*)]^2}{(a_P/R_*)^2 \varrho_{P,T}^2 - [b_{P,T}+(\Delta Y_P/R_*)]^2}} \tag{6.74}$$

where the "TIP" superscript indicates that this definition of the duration only accounts for the TDV-TIP effect and not the TDV-V. Since the V component is just a scaling factor, it can be applied *after* the TIP component has been incorporated. This two-step process allows one to write the transit duration accounting for both V and TIP components as:

$$\tilde{T}_P = \frac{\dot{X}_B}{\dot{X}_B + \Delta \dot{X}_P} \tilde{T}_P^{\mathrm{TIP}} \tag{6.75}$$

At this point, it is useful to define $\tilde{T}_P^{\text{TIP}}$ as some factor multiplied by the barycentric transit duration, $\tilde{T}_B$:

$$\epsilon = \frac{\tilde{T}_P^{\text{TIP}}}{\tilde{T}_B} \tag{6.76}$$

Note that $\tilde{T}$ increases as the impact parameter approaches zero. Since Y_B is generally positive, then increases in the transit duration occur when ΔY_P is negative and decreases occur when ΔY_P is positive. Mathematically, $\epsilon > 1$ for $\Delta Y_P < 0$ and $\epsilon < 1$ for $\Delta Y_P > 0$. The total TDV effect may be now written as:

$$\text{TDV} = \left[\frac{\epsilon \dot{X}_B}{\dot{X}_B + \Delta \dot{X}_P} - 1\right] \tilde{T}_B \tag{6.77}$$

Using assumption $\alpha 7$, this simplifies to:

$$\text{TDV} = \left[(\epsilon - 1) - \frac{\epsilon \Delta \dot{X}_P}{\dot{X}_B}\right] \tilde{T}_B \tag{6.78}$$

One can now see that the TDV signal has two clear components. In the absence of any TIP-component, $\epsilon \to 1$ and one recovers the TDV-V effect seen earlier. Given that ϵ is of order unity, it is useful to write it as:

$$\epsilon = 1 + \varphi \tag{6.79}$$

where it is understood that φ is small compared to unity. Expanding out, the TDV effect now becomes:

$$\text{TDV} = \left[\varphi - \frac{\Delta \dot{X}_P}{\dot{X}_B} - \frac{\varphi \Delta \dot{X}_P}{\dot{X}_B}\right] \tilde{T}_B \tag{6.80}$$

Both $(\Delta \dot{X}_P / \dot{X}_B)$ and φ are small terms and thus the cross-product of the two gives an even smaller term i.e. it is second order. Therefore, the total TTV effect may be expressed as:

$$\begin{aligned} \text{TDV} &= \left[\varphi - \frac{\Delta \dot{X}_P}{\dot{X}_B}\right] \tilde{T}_B \\ \text{TDV} &= (\text{TDV-TIP}) + (\text{TDV-V}) \end{aligned} \tag{6.81}$$

$$\text{TDV-TIP} = \varphi \tilde{T}_B \tag{6.82}$$

Critically, TDV may be treated as a linear combination of the TDV-V and TDV-TIP components, which grossly simplifies the subsequent analysis.

6.6.2 Derivation

To provide the TDV-TIP waveform and RMS amplitude, it is necessary to derive φ, which is a function of ΔY_P. ϵ is given by:

$$\epsilon = \frac{1}{\tilde{T}_B}\frac{P_P}{\pi}\frac{\varrho_{P,T}^2}{\sqrt{1-e_P^2}}\arcsin\sqrt{\frac{1-[b_{P,T}+(\Delta Y_P/R_*)]^2}{(a_P/R_*)^2\varrho_{P,T}^2-[b_{P,T}+(\Delta Y_P/R_*)]^2}} \tag{6.83}$$

Assuming $(a_P/R_*)^2\varrho_{P,T}^2 \gg b_{P,T}^2$ and using a small-angle approximation, ϵ may be written as:

$$\epsilon^2 = \frac{1-[b_{P,T}+(\Delta Y_P/R_*)]^2}{1-[b_{P,T}]^2} \tag{6.84}$$

Note that the above is true even for eccentric orbits. Expanding out the squares:

$$\begin{aligned}\epsilon^2 &= \frac{1-b_{P,T}^2-2b_{P,T}(\Delta Y_P/R_*)-\mathcal{O}[(\Delta Y_P/R_*)^2]}{1-b_{P,T}^2}\\ \epsilon &= \sqrt{1-\frac{2b_{P,T}(\Delta Y_P/R_*)}{1-b_{P,T}^2}}\end{aligned} \tag{6.85}$$

where the second line has made the reasonable approximation that $\mathcal{O}[(\Delta Y_P/R_*)^2]$ is small, since it is second order. Since one knows that ϵ is of order unity due to its definition as the ratio of the transit duration with and without the TDV effect, then the term subtracted from unity inside the square root must be small i.e. $\varphi \ll 1$. Therefore, a Taylor expansion to first-order gives:

$$\varphi \simeq -\frac{b_{P,T}(\Delta Y_P/R_*)}{1-b_{P,T}^2} \tag{6.86}$$

This tells us that for $\Delta Y_P > 0$, $\varphi < 0$ indicating $\epsilon < 1$ which in turn means the duration has decreased, which follows the expected behaviour. The TDV-TIP waveform is therefore:

$$\begin{aligned}\text{TDV-TIP} &= \tilde{T}_B\left(\frac{b_{P,T}}{1-b_{P,T}^2}\right)\left(\frac{a_S M_S(1-e_S^2)}{R_* M_P}\right)\Lambda_{\text{TDV-TIP}}\\ \Lambda_{\text{TDV-TIP}} &= \\ &\frac{\sin(f_S+\omega_S)[-\cos i_S\sin i_P+\sin i_S\cos i_P\cos\varpi_S]+\cos i_P\sin\varpi_S\cos(\omega_S+f_S)}{1+e_S\cos f_S}\end{aligned} \tag{6.87}$$

In [16], I had been unable to solve the integral of $\Lambda_{\text{TDV-TIP}}^2$, but I here present a solution found through a decomposition of the various elements, integrated and then re-combined:

$$\delta_{\text{TDV-TIP}} = \tilde{T}_B \left(\frac{b_{P,T}}{1 - b_{P,T}^2} \right) \left(\frac{a_S M_S (1 - e_S^2)}{R_* M_P} \right) \sqrt{\frac{\Phi_{\text{TDV-TIP}}}{2\pi}}$$

$$\begin{aligned}
\Phi_{\text{TDV-TIP}} = \pi e_S^{-2} \left(1 - e_S^2\right)^{-3/2} \Bigg[& - \left(-e_S^2 + \left(e_S^2 \left(3 - 2\sqrt{1 - e_S^2} \right) \right. \right. \\
& \left. \left. + 2(-1 + \sqrt{1 - e_S^2}) \right) \cos 2\omega_S \right) \\
& \times [\cos i_S \sin i_P - \cos i_P \cos \varpi_S \sin i_S]^2 \\
& + 2 \left(2 - 2\sqrt{1 - e_S^2} + e_S^2 \left(-3 + 2\sqrt{1 - e_S^2} \right) \right) \cos i_P \sin 2\omega_S \sin \varpi_S \\
& \times [-\cos i_S \sin i_P + \cos i_P \cos \varpi_S \sin i_S] \\
& - \cos^2 i_P \sin^2 \varpi_S \left(-e_S^2 + \cos 2\omega_S \right. \\
& \left. \times \left(2 - 2\sqrt{1 - e_S^2} + e_S^2 \left(-3 + 2\sqrt{1 - e_S^2} \right) \right) \right) \Bigg]
\end{aligned} \tag{6.88}$$

In astrophysical units, this becomes:

$$\begin{aligned}
\delta_{\text{TDV-TIP}} = (7.73\,\text{s})\, \mathfrak{D}^{1/2} & \left(\frac{\tilde{T}_B}{10\,\text{h}} \right) \left(\frac{b_{P,T}}{1 - b_{P,T}^2} \right) \left(\frac{a_P}{R_*} \right) \left(\frac{M_S}{M_\oplus} \right) \left(\frac{M_J}{M_P} \right)^{2/3} \\
& \times \left(\frac{M_\odot}{M_*} \right)^{1/3} (1 - e_S^2) \sqrt{\frac{\Phi_{\text{TDV-TIP}}}{2\pi}}
\end{aligned} \tag{6.89}$$

6.6.3 Properties of the TDV-TIP RMS Amplitude

As with the other Φ terms, $\Phi_{\text{TDV-TIP}} \to \pi$ in the nominal case. Unlike, the other Φ terms though, this nominal case occurs for $i_S = 0$ rather than $i_S = \pi/2$ since this effect is non-coplanar, whereas the others were coplanar. Taking the limit for circular orbits and $\varpi_S = 0$, $\Phi_{\text{TDV-TIP}}$ becomes:

$$\lim_{e_S \to 0} \lim_{\varpi_S \to 0} \Phi_{\text{TDV-TIP}} = \pi \sin^2(i_P - i_S) \tag{6.90}$$

This indicates that maximum TIP effects occur for $i_P - i_S = n(\pi/2)$, where n is any real integer. This prediction is exactly what one would expect intuitionally, as it corresponds to the planet's reflex motion being orthogonal to the $\hat{X}$-direction. For the eccentricity dependency, I take the limit for $i_S = 0$, $\varpi_S = 0$ and $i_P = \pi/2$ and then one may plot $\Phi_{\text{TDV-TIP}}$ purely as a function of e_S and ω_S, as shown in Fig. 6.6.

Another interesting point to bear in mind with TDV-TIP is that one does not require the moon to be inclined relative to the planet's orbital plane to still have a TIP effect. The only requirement is that $i_P - i_S \neq n\pi$. Therefore, if $i_P \neq \pi/2$,

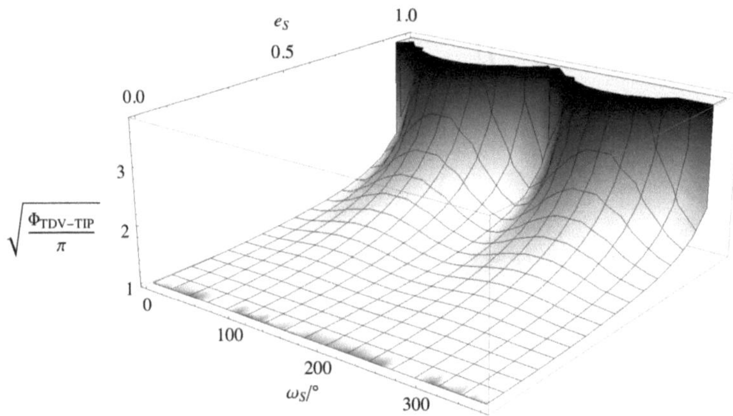

Fig. 6.6 Dependency of $\Phi_{\text{TDV-TIP}}$, the TDV-TIP scaling factor, as a function of the satellite's eccentricity and argument of periapsis for $i_P = \pi/2$ and $i_S = 0$

then a TIP effect is guaranteed unless one has the coincidence that $i_S = n\pi + i_P$. Consider the case of $e_S \ll 1$, and $i_S = \pi/2$, the dependency on ϖ_S naturally drops out leaving one with:

$$\lim_{e_S \to 0} \lim_{i_S \to \pi/2} \Phi_{\text{TDV-TIP}} = \pi \cos^2 i_P \tag{6.91}$$

This last equation is particularly useful because one expects moons to exhibit circular, coplanar configurations based upon the contracted region of orbital stability [18] but there is no reason why the planet's orbital plane should coincidentally align to the observer's line-of-sight.

6.6.4 The Total TDV Effect

As shown in Sect. 6.6.1, the total TDV effect may be treated as a linear combination of the two components. This yields:

$$\begin{aligned} \text{TDV} = \tilde{T}_B \left(\frac{a_S M_S P_P}{a_P M_P P_S} \right) \left(\frac{\sqrt{1-e_P^2}}{(1+e_P \sin\omega_P)\sqrt{1-e_S^2}} \right) \Lambda_{\text{TDV-V}} \\ + \tilde{T}_B \left(\frac{b_{P,T}}{1-b_{P,T}^2} \right) \left(\frac{a_S M_S (1-e_S^2)}{R_* M_P} \right) \Lambda_{\text{TDV-TIP}} \end{aligned} \tag{6.92}$$

The RMS amplitude integral will be given by:

$$2\pi\delta_{\text{TDV}}^2 = \int_{f_S=0}^{2\pi} \tilde{T}_B^2 \left(\frac{a_S M_S P_P}{a_P M_P P_S}\right)^2 \left(\frac{\sqrt{1-e_P^2}}{(1+e_P\sin\omega_P)\sqrt{1-e_S^2}}\right)^2 \Lambda_{\text{TDV-V}}^2$$
$$+ 2\tilde{T}_B^2 \left(\frac{a_S M_S P_P}{a_P M_P P_S}\right)\left(\frac{\sqrt{1-e_P^2}}{(1+e_P\sin\omega_P)\sqrt{1-e_S^2}}\right)$$
$$\times \left(\frac{b_{P,T}}{1-b_{P,T}^2}\right)\left(\frac{a_S M_S(1-e_S^2)}{R_* M_P}\right)\Lambda_{\text{TDV-V}}\Lambda_{\text{TDV-TIP}}$$
$$+ \tilde{T}_B^2\left(\frac{b_{P,T}}{1-b_{P,T}^2}\right)^2\left(\frac{a_S M_S(1-e_S^2)}{R_* M_P}\right)^2 \Lambda_{\text{TDV-TIP}}^2 \,\mathrm{d}f_S \qquad (6.93)$$

The cross-term is new, but the two squared terms were integrated previously:

$$2\pi\delta_{\text{TDV}}^2 = \tilde{T}_B^2 \left(\frac{a_S M_S P_P}{a_P M_P P_S}\right)^2 \left(\frac{\sqrt{1-e_P^2}}{(1+e_P\sin\omega_P)\sqrt{1-e_S^2}}\right)^2 \Phi_{\text{TDV-V}}$$
$$+ 2\tilde{T}_B^2\left(\frac{a_S^2 M_S^2 P_P}{R_* a_P M_P^2 P_S}\right)\left(\frac{\sqrt{1-e_P^2}\sqrt{1-e_S^2}}{(1+e_P\sin\omega_P)}\right)\left(\frac{b_{P,T}}{1-b_{P,T}^2}\right)\Phi_{\text{cross}}$$
$$+ \tilde{T}_B^2\left(\frac{b_{P,T}}{1-b_{P,T}^2}\right)^2\left(\frac{a_S M_S(1-e_S^2)}{R_* M_P}\right)^2 \Phi_{\text{TDV-TIP}} \qquad (6.94)$$

where the Φ_{cross} term is given by:

$$\Phi_{\text{cross}} = \int_{f_S=0}^{2\pi} \Lambda_{\text{TDV-V}}\Lambda_{\text{TDV-TIP}}\,\mathrm{d}f_S$$

$$\Phi_{\text{cross}} =$$
$$-\frac{\pi}{e_S^2}\Bigg[\frac{\cos i_P \sin i_S \sin^2\varpi_S\left[2e_S^2\cos^2\omega_S\left(\sqrt{1-e_S^2}-1\right)-\cos 2\omega_S\left(e_S^2+2\sqrt{1-e_S^2}-2\right)+e_S^2\right]}{\sqrt{1-e_S^2}}$$
$$+\sin i_S \sin 2\omega_S \sin\varpi_S[\cos i_P \sin i_S \cos\varpi_S - \sin i_P \cos i_S]\left[e_S^2+2\sqrt{1-e_S^2}-2\right]$$
$$+\cos 2\varpi_S[\cos i_P \sin i_S \cos\varpi_S - \sin i_P \cos i_S]\left[e_S^2-\cos 2\omega_S\left(e_S^2+2\sqrt{1-e_S^2}-2\right)\right]$$
$$+\frac{1}{2}\cos i_P \sin 2\omega_S \sin 2\varpi_S\left[e_S^2+2\sqrt{1-e_S^2}-2\right]\Bigg] \qquad (6.95)$$

The cross-term, whose "parents" could be considered to be $\Phi_{\text{TDV-V}}$ and $\Phi_{\text{TDV-TIP}}$, mimics the properties of the $\Phi_{\text{TDV-TIP}}$ parent in the sense of its nominal case. Like $\Phi_{\text{TDV-TIP}}$, $\Phi_{\text{cross}} \to \pi$ when $i_S \to 0$ rather than $\pi/2$. Therefore, the cross-term describes an out-of-the-plane effect. This is supported by the fact $\Phi_{\text{TDV-TIP}} \to 0$ when $i_S \to 0$ for a coplanar planet. One useful limiting case to consider is that of a circular moon, which yields:

$$\lim_{e_S \to 0} \Phi_{\text{cross}} = \pi \left(\cos i_S \sin i_P \cos \varpi_S - \cos i_P \sin i_S \right) \tag{6.96}$$

6.6.5 *The Consequences for η*

η, as defined in (6.69), is the ratio of the TDV to TTV RMS amplitudes and was shown earlier to allow one to directly determine P_S and then M_S, thus solving Problem 1 in Sect. 6.2.4. With the TDV-TIP effect now recognized, the expression for η will become more complicated.

Equation 6.94 presented the total TDV RMS amplitude as the quadrature sum of the two squares of the individual components plus a cross-term. It is useful to complete the square in (6.94). If one defines $\mathcal{W}_{\text{TDV-V}} = \text{TDV-V}/\Lambda_{\text{TDV-V}}$ and similarly for the TDV-TIP effect, then one may write:

$$\begin{aligned} 2\pi \delta_{\text{TDV}}^2 &= \mathcal{W}_{\text{TDV-V}}^2 \Phi_{\text{TDV-V}}^2 + \mathcal{W}_{\text{TDV-TIP}}^2 \Phi_{\text{TDV-TIP}}^2 + 2\mathcal{W}_{\text{TDV-V}} \mathcal{W}_{\text{TDV-TIP}} \Phi_{\text{cross}} \\ &= [\mathcal{W}_{\text{TDV-V}} \Phi_{\text{TDV-V}} + \mathcal{W}_{\text{TDV-TIP}} \Phi_{\text{TDV-TIP}}]^2 + 2\mathcal{W}_{\text{TDV-V}} \mathcal{W}_{\text{TDV-TIP}} \\ &\quad \times [\Phi_{\text{cross}} - \Phi_{\text{TDV-V}} \Phi_{\text{TDV-TIP}}] \\ &= [\mathcal{W}_{\text{TDV-V}} \Phi_{\text{TDV-V}} + \mathcal{W}_{\text{TDV-TIP}} \Phi_{\text{TDV-TIP}}]^2 + 2\mathcal{W}_{\text{TDV-V}} \mathcal{W}_{\text{TDV-TIP}} \Delta\Phi \end{aligned} \tag{6.97}$$

where I have defined $\Delta\Phi = \Phi_{\text{cross}} - \Phi_{\text{TDV-V}} \Phi_{\text{TDV-TIP}}$. In order for the squares and square roots to cancel, one ideally wants $\Delta\Phi$ to be small. Assuming $e_S \ll 1$ and $i_S \simeq \pi/2$ in line with the dynamical expectations [18], $\Delta\Phi$ is given by:

$$\lim_{e_S \to 0} \lim_{i_S \to \pi/2} \lim_{\varpi_S \to 0} \Delta\Phi = \pi \cos i_P + \mathcal{O}[\cos^2 i_P] \tag{6.98}$$

Since the other Φ terms are of order unity and $\cos i_P \ll 1$, then in general, one can see that the $\Delta\Phi$ term will be negligible. In such a circumstance, the total TDV amplitude becomes:

$$\begin{aligned} \delta_{\text{TDV}} \simeq \tilde{T}_B \left(\frac{a_S M_S P_P}{a_P M_P P_S} \right) \left(\frac{\sqrt{1 - e_P^2}}{(1 + e_P \sin \omega_P)\sqrt{1 - e_S^2}} \right) \sqrt{\frac{\Phi_{\text{TDV-V}}}{2\pi}} \\ + \tilde{T}_B \left(\frac{b_{P,T}}{1 - b_{P,T}^2} \right) \left(\frac{a_S M_S (1 - e_S^2)}{R_* M_P} \right) \sqrt{\frac{\Phi_{\text{TDV-TIP}}}{2\pi}} \end{aligned} \tag{6.99}$$

Table 6.1 Summary of key properties of the three known transit timing effects due to an exomoon

	TTV	TDV-V	TDV-TIP
Type of effect	Positional	Velocity	Positional
Direction	$\hat{X}$	$\hat{X}$	$\hat{Y}$
Proportionality	$M_S a_S$	$M_S a_S^{-1/2}$	$M_S a_S$
Relative phase	0	$\pi/2$	$\pm\pi/2$
Waveform equation	Eq. 6.44	Eq. 6.63	Eq. 6.87
RMS amplitude equation	Eq. 6.46	Eq. 6.65	Eq. 6.88

The expression for η now becomes:

$$\begin{aligned}\eta &= \frac{\delta_{\mathrm{TDV}}}{\delta_{\mathrm{TTV}}} \\ &= \frac{\tilde{T}_B}{P_S}\frac{1}{\left(1-e_S^2\right)^{3/2}}\sqrt{\frac{\Phi_{\mathrm{TDV\text{-}V}}}{\Phi_{\mathrm{TTV}}}} \\ &\quad + \frac{\tilde{T}_B}{P_B}\frac{a_P}{R_*}\left(\frac{b_{P,T}}{1-b_{P,T}^2}\right)\left(\frac{1+e_P\sin\omega_P}{\sqrt{1-e_P^2}}\right)\sqrt{\frac{\Phi_{\mathrm{TDV\text{-}TIP}}}{\Phi_{\mathrm{TTV}}}}\end{aligned} \tag{6.100}$$

Now consider the case of $e_S \ll 1$ and $i_S \simeq \pi/2$, as was done previously:

$$\begin{aligned}\lim_{e_S\to 0}\lim_{i_S\to\pi/2}\eta &= \frac{\tilde{T}_B}{P_S}\sqrt{\frac{\Phi_{\mathrm{TDV\text{-}V}}}{\Phi_{\mathrm{TTV}}}} + \frac{\tilde{T}_B}{P_B}\frac{a_P}{R_*}\left(\frac{b_{P,T}}{1-b_{P,T}^2}\right)\sqrt{\frac{\Phi_{\mathrm{TDV\text{-}TIP}}}{\Phi_{\mathrm{TTV}}}} \\ &= \frac{\tilde{T}_B}{P_S} + \frac{\tilde{T}_B}{P_B}\left(\frac{b_{P,T}^2}{1-b_{P,T}^2}\right)\end{aligned} \tag{6.101}$$

Therefore, the contribution of the TIP effect is just to add a constant onto the η term. This constant is purely a function of the properties of the planet, and so can be estimated reliably.

A summary of the properties of TTV, TDV-V and TDV-TIP is provided in Table 6.1 and I provide some estimates of the TTV and TDV amplitudes for several transiting exoplanet systems in Table 6.2.

6.6.6 Prograde Versus Retrograde

It was shown earlier in Sect. 6.5.2 how a retrograde orbit does not alter the phase shift between TTV and TDV-V. Retrograde orbits are defined by $\pi < i_S < 2\pi$, or alternatively as when the sense of the moon's orbital motion around the planet is counter to that of the planet around the star. A retrograde orbit would indicate a capture origin for the exomoon and thus its measurement would be a major milestone in the road to understanding satellite formation outside of the Solar System. The TDV-TIP effect offers a way to make this determination.

Table 6.2 Predicted TTV and TDV (both V- and TIP-components) RMS amplitudes due to a 1 $M_{\oplus}$ exomoon at 1/3 the Hill radius, for a selection of candidate transiting planets

Planet	δ_{TTV}/s	$\delta_{TDV\text{-}V}$/s	$\delta_{TDV\text{-}TIP}$/s	Reference
HAT-P-11b	19.19	22.54	0.40	[19]
GJ 436b	14.12	13.68	1.30	[20]
CoRoT-4b	7.58	9.15	0.00	[21]
OGLE-TR-111b	4.63	7.32	0.11	[22]
HAT-P-1b	4.58	6.82	0.47	[23]
HD 149026b	3.61	9.76	0.00	[24]
Lupus-TR-3b	3.28	5.19	0.07	[25]
WASP-7b	3.26	5.88	0.00	[26]
HD 17156b	3.07	1.06	0.43	[27]
TrES-1b	3.04	5.95	0.05	[28]
HD2 09458b	2.97	5.95	0.07	[7]
XO-5b	2.65	4.69	0.17	[29]
HAT-P-4b	2.54	8.34	0.00	[30]
HD 189733b	1.52	2.96	0.16	[31]
XO-3b	0.41	0.87	0.07	[32]

System parameters are taken from various references, which are shown

Consider the planet–moon barycentre moving in the $+\hat{X}$-direction in Fig. 6.5. For a prograde orbit, the velocity of the planet around the planet–moon barycentre must be in the $+\hat{X}$-direction when it is at position (1). At position (1), the transit impact parameter has increased and thus the transit duration has shortened. At the same time, the planet's reflex velocity is additive to the planet–moon barycentre velocity around the host star, and so the transit duration is further shortened. Thus for prograde orbits, it can be seen that the TIP- and V-components are additive. The reverse logic is true for retrograde orbits.

Whilst this intuitive explanation is useful, the validation may be provided by considering the Λ terms for each timing effect. For the conditions used earlier in Sect. 6.5.2, a prograde moon has:

$$\begin{aligned}
\lim_{e_P,e_S\to 0}\lim_{\Omega_S\to 0}\lim_{i_S\to\pi/2}\Lambda_{\text{TTV}} &= \cos f_S \\
\lim_{e_P,e_S\to 0}\lim_{\Omega_S\to 0}\lim_{i_S\to\pi/2}\Lambda_{\text{TDV-V}} &= \sin f_S \\
\lim_{e_P,e_S\to 0}\lim_{\Omega_S\to 0}\lim_{i_S\to\pi/2}\Lambda_{\text{TDV-TIP}} &= \sin f_S \cos i_P
\end{aligned} \tag{6.102}$$

But for a retrograde moon, this becomes:

$$\begin{aligned}
\lim_{e_P,e_S\to 0}\lim_{\Omega_S\to 0}\lim_{i_S\to 3\pi/2}\Lambda_{\text{TTV}} &= \cos f_S \\
\lim_{e_P,e_S\to 0}\lim_{\Omega_S\to 0}\lim_{i_S\to 3\pi/2}\Lambda_{\text{TDV-V}} &= \sin f_S \\
\lim_{e_P,e_S\to 0}\lim_{\Omega_S\to 0}\lim_{i_S\to 3\pi/2}\Lambda_{\text{TDV-TIP}} &= -\sin f_S \cos i_P
\end{aligned} \tag{6.103}$$

So the TDV-TIP effect can be seen to flip its phase shift. For prograde orbits, the TDV-V and TDV-TIP effects are in phase and thus constructively interfere. For a retrograde orbit, destructive interference occurs. If one assumes that in general the TDV-V component is larger than the TDV-TIP component, which can be justified by the typical examples given in Table 6.2, then for a retrograde orbit, the constant term in η becomes negative rather than positive:

$$\lim_{e_S \to 0} \lim_{i_S \to \pi/2} \eta = \frac{\tilde{T}_B}{P_S} \pm \frac{\tilde{T}_B}{P_B} \left(\frac{b_{P,T}^2}{1 - b_{P,T}^2} \right) \tag{6.104}$$

Since the constant term is controlled only by the planetary properties, this raises the possibility of measuring the sense of orbital motion. For any set of measurements, two versions of η can be constructed, one for prograde moons and one for retrograde. The orbital period is derived using η and then this orbital period could be checked against the periodogram for both signals. The version of η which provides the closest agreement to a periodogram peak would be accepted as the real one, and thus the sense of orbital motion could be determined (or at least some odds ratio of prograde to retrograde).

6.7 Conclusions

In this chapter, I have provided a self-consistent model for the transit timing effects due to an extrasolar moon. The approximations and assumptions have been carefully considered and outlined, yet the model maintains a full three-dimensional framework including the effects of orbital eccentricity. The predicted effects are transit timing variations (TTV), velocity induced transit duration variations (TDV-V) and transit impact parameter induced transit duration variations (TDV-TIP).

TTV and TDV-TIP are due to the changes in position of the planet whereas TDV-V is a velocity effect. TTV by itself provides $M_S a_S$ only due to sampling constraints, but combining TTV with TDV-V breaks the mass degeneracy and allows one to measure M_S and a_S separately. Further, the two effects exhibit a $\pi/2$ shift allowing one to unambiguously identify the signals as being due to an exomoon, solving another long-standing problem with TTV alone.

I have also discussed non-coplanar effects, notably TDV-TIP, which is due to kicks out of the plane. TDV-TIP is generally an order-of-magnitude (or more) smaller than the TDV-V component (e.g. Table 6.2) and so is usually a second-order effect. However, even for moons which are aligned to the planet's orbital plane, the effect persists but fortunately it does not invalidate the mass and period determination for the moon, it merely complicates the analysis somewhat. In fact, combining the TDV to TTV amplitude ratio with some harmonic information from the periodogram allows one to determine the orbital sense of motion of the moon. Naturally, this would require a large signal to noise in the timing effects though.

The final expressions for the RMS amplitudes of the TTV, TDV-V and TDV-TIP effects are given in (6.46), (6.65) and (6.88) respectively and the combined TDV RMS amplitude in (6.99). Also see Table 6.1 for a concise summary.

This chapter completes the theoretical aspect of TTV and TDV, but one question which remains is how feasible a search for exomoons using these techniques would actually be. In Chap. 7, I will present a feasibility study using the recently launched *Kepler Mission*.

References

1. P. Sartoretti, J. Schneider, On the detection of satellites of extrasolar planets with the method of transits. Astron. Astrophys. Suppl. **134**, 553–560 (1999)
2. E. Agol, J. Steffen, R. Sari, W. Clarkson, On detecting terrestrial planets with timing of giant planet transits. Mon. Not. R. Astron. Soc. **359**, 567–579 (2005)
3. M.J. Holman, N.W. Murray, The use of transit timing to detect terrestrial-mass extrasolar planets. Science **307**, 1288–1291 (2005)
4. M.J. Holman, D.C. Fabrycky, D. Ragozzine, E.B. Ford, J.H. Steffen, W.F. Welsh, J.J. Lissauer, D.W. Latham, G.W. Marcy, L.M. Walkowicz, N.M. Batalha, J.M. Jenkins, J.F. Rowe, W.D. Cochran, F. Fressin, G. Torres, L.A. Buchhave, D.D. Sasselov, W.J. Borucki, D.G. Koch, G. Basri, T.M. Brown, D.A. Caldwell, D. Charbonneau, E.W. Dunham, T.N. Gautier, J.C. Geary, R.L. Gilliland, M.R. Haas, S.B. Howell, D.R. Ciardi, M. Endl, D. Fischer, G. Fürész, J.D. Hartman, H. Isaacson, J.A. Johnson, P.J. MacQueen, A.V. Moorhead, R.C. Morehead, J.A. Orosz, Kepler-9: a system of multiple planets transiting a sun-like star, confirmed by timing variations. Science **330**, 51 (2010)
5. E.B. Ford, M.J. Holman, Using transit timing observations to search for trojans of transiting extrasolar planets. Astrophys. J. Lett. **664**, L51–L54 (2007)
6. C.A. Scharf, Exoplanet transit parallax. Astrophys. J. **661**, 1218–1221 (2007)
7. D.M. Kipping, Transiting planets—light-curve analysis for eccentric orbits. Mon. Not. R. Astron. Soc. **389**, 1383–1390 (2008)
8. A. Jordán, G.Á. Bakos, Observability of the general relativistic precession of periastra in exoplanets. Astrophys. J. **685**, 543–552 (2008)
9. R. Alonso, M. Auvergne, A. Baglin, M. Ollivier, C. Moutou, D. Rouan, H.J. Deeg, S. Aigrain, J.M. Almenara, M. Barbieri, P. Barge, W. Benz, P. Bordé, F. Bouchy, de La R. Reza, M. Deleuil, R. Dvorak, A. Erikson, M. Fridlund, M. Gillon, P. Gondoin, T. Guillot, A. Hatzes, G. Hébrard, P. Kabath, L. Jorda, H. Lammer, A. Léger, A. Llebaria, B. Loeillet, P. Magain, M. Mayor, T. Mazeh, M. Pätzold, F. Pepe, F. Pont, D. Queloz, H. Rauer, A. Shporer, J. Schneider, B. Stecklum, S. Udry, G. Wuchterl, Transiting exoplanets from the CoRoT space mission. II. CoRoT-Exo-2b: a transiting planet around an active G star. Astron. Astrophys. **482**, L21–L24 (2008)
10. R.R. Rafikov, Stellar Proper Motion and the Timing of Planetary Transits. Astrophys. J. **700**, 965–970 (2009)
11. C. Hellier, D.R. Anderson, A.C. Cameron, M. Gillon, L. Hebb, P.F.L. Maxted, D. Queloz, B. Smalley, A.H.M.J. Triaud, R.G. West, D.M. Wilson, S.J. Bentley, B. Enoch, K. Horne, J. Irwin, T.A. Lister, M. Mayor, N. Parley, F. Pepe, D.L. Pollacco, D. Segransan, S. Udry, P.J. Wheatley, An orbital period of 0.94days for the hot-Jupiter planet WASP-18b. Nature **460**, 1098–1100 (2009)
12. C.A. Watson, T.R. Marsh, Orbital period variations of hot Jupiters caused by the Applegate effect. Mon. Not. R. Astron. Soc. **405**, 2037–2043 (2010)

13. M. Montalto, Planetary transit timing variations induced by stellar binarity. The light travel time effect. Astron. Astrophys. **521**, A60+ (2010)
14. R.C. Domingos, O.C. Winter, T. Yokoyama, Stable satellites around extrasolar giant planets. Mon. Not. R. Astron. Soc. **373**, 1227–1234 (2006)
15. D.M. Kipping, Transit timing effects due to an exomoon. Mon. Not. R. Astron. Soc. **392**, 181–189 (2009)
16. D.M. Kipping, Transit timing effects due to an exomoon—II. Mon. Not. R. Astron. Soc. **396**, 1797–1804 (2009)
17. C.D. Murray, S.F. Dermott, Solar system dynamics. (Cambridge University Press, Cambridge, 1999)
18. J.R. Donnison, The Hill stability of inclined small mass binary systems in three-body systems with special application to triple star systems, extrasolar planetary systems and Binary Kuiper Belt systems. Planet. Space Sci. **58**:1169–1179, August (2010). doi:10.1016/j.pss.2010.04.009
19. D.M. Kipping, G.Á. Bakos, Analysis of Kepler's short-cadence photometry for TrES-2b. ArXiv e-prints, June 2010
20. G. Torres, J.N. Winn, M.J. Holman, Improved parameters for extrasolar transiting planets. Astrophys. J. **677**, 1324–1342 (2008)
21. S. Aigrain, A. Collier Cameron, M. Ollivier, F. Pont, L. Jorda, J.M. Almenara, R. Alonso, P. Barge, P. Bordé, F. Bouchy, H. Deeg, de La R. Reza, M. Deleuil, R. Dvorak, A. Erikson, M. Fridlund, P. Gondoin, M. Gillon, T. Guillot, A. Hatzes, H. Lammer, A.F. Lanza, A. Léger, A. Llebaria, P. Magain, T. Mazeh, C. Moutou, M. Paetzold, C. Pinte, D. Queloz, H. Rauer, D. Rouan, J. Schneider, G. Wuchter, S. Zucker, Transiting exoplanets from the CoRoT space mission. IV. CoRoT-Exo-4b: a transiting planet in a 9.2 day synchronous orbit. Astron. Astrophys. **488**, L43–L46 (2008)
22. R.F. Díaz, P. Rojo, M. Melita, S. Hoyer, D. Minniti, P.J.D. Mauas, M.T. Ruíz, Detection of period variations in extrasolar transiting planet OGLE-TR-111b. Astrophys. J. Lett. **682**, L49–L52 (2008)
23. J.A. Johnson, J.N. Winn, N. Narita, K. Enya, P.K.G. Williams, G.W. Marcy, B. Sato, Y. Ohta, A. Taruya, Y. Suto, E.L. Turner, G. Bakos, R.P. Butler, S.S. Vogt, W. Aoki, M. Tamura, T. Yamada, Y. Yoshii, M. Hidas Measurement of the spin–orbit angle of exoplanet HAT-P-1b. Astrophys. J. **686**:649–657, October 2008. doi:10.1086/591078
24. J.N. Winn, G.W. Henry, G. Torres, M.J. Holman, Five new transits of the Super-Neptune HD 149026b. Astrophys. J. **675**, 1531–1537 (2008)
25. D.T.F. Weldrake, D.D.R. Bayliss, P.D. Sackett, B.W. Tingley, M. Gillon, J. Setiawan, Lupus-TR-3b: a low-mass transiting hot Jupiter in the Galactic Plane?. Astrophys. J. Lett. **675**, L37–L40 (2008)
26. C. Hellier, D.R. Anderson, M. Gillon, T.A. Lister, P.F.L. Maxted, D. Queloz, B. Smalley, A.H.M.J. Triaud, R.G. West, D.M. Wilson, K. Alsubai, S.J. Bentley, A.C. Cameron, L. Hebb, K. Horne, J. Irwin, S.R. Kane, M. Mayor, F. Pepe, D. Pollacco, I. Skillen, S. Udry, P.J. Wheatley, D.J. Christian, R. Enoch, C.A. Haswell, Y.C. Joshi, A.J. Norton, N. Parley, R. Ryans, R.A. Street, I. Todd, Wasp-7: a bright transiting-exoplanet system in the Southern Hemisphere. Astrophys. J. Lett. **690**, L89–L91 (2009)
27. M. Barbieri, R. Alonso, G. Laughlin, J.M. Almenara, R. Bissinger, D. Davies, D. Gasparri, E. Guido, C. Lopresti, F. Manzini, G. Sostero, HD 17156b: a transiting planet with a 21.2-day period and an eccentric orbit. Astron. Astrophys. **476**, L13–L16 (2007)
28. J.N. Winn, M.J. Holman, A. Roussanova, The transit light curve project. III. Tres transits of TrES-1. Astrophys. J. **657**, 1098–1106 (2007)
29. C.J. Burke, P.R. McCullough, J.A. Valenti, D. Long, C.M. Johns-Krull, P. Machalek, K.A. Janes, B. Taylor, M.L. Fleenor, C.N. Foote, B.L. Gary, E. García-Melendo, J. Gregorio, T. Vanmunster, XO-5b: a transiting Jupiter-sized planet with a 4 day period. Astrophys. J. **686**, 1331–1340 (2008)
30. G. Kovács, G.Á. Bakos, G. Torres, A. Sozzetti, D.W. Latham, R.W. Noyes, R.P. Butler, G.W. Marcy, D.A. Fischer, J.M. Fernández, G. Esquerdo, D.D. Sasselov, R.P. Stefanik, A. Pál, J.

Lázár, I. Papp, P. Sári, HAT-P-4b: a metal-rich low-density transiting hot Jupiter. Astrophys. J. Lett. **670**, L41–L44 (2007)
31. J.P. Beaulieu, S. Carey, I. Ribas, G. Tinetti, Primary transit of the planet HD 189733b at 3.6 and 5.8 μm.. Astrophys. J. **677**, 1343–1347 (2008)
32. J.N. Winn, M.J. Holman, G. Torres, P. McCullough, C. Johns-Krull, D.W. Latham, A. Shporer, T. Mazeh, E. Garcia-Melendo, C. Foote, G. Esquerdo, M. Everett, The transit light curve project. IX. Evidence for a smaller radius of the exoplanet XO-3b. Astrophys. J. **683**, 1076–1084 (2008)

Chapter 7
Detectability of Habitable Exomoons with *Kepler*-Class Photometry

Imagination will often carry us to worlds that never were. But without it we go nowhere

Carl Sagan, Cosmos, 1980

7.1 Introduction

In Sect. 6.1, I discussed how extrasolar moons may be detected around transiting planets by searching for timing deviations in both transit minima and duration (TTV and TDV respectively). Whilst the theory and modelling of these effects has been covered in detail in the previous chapter, there has been little discussion of how feasible this enterprise actually is.

The goal of this chapter will be to provide quantitative estimates of the minimum exomoon mass which can be detected with current technology and facilities. To measure these small timing deviations, one requires a highly precise photometric instrument with uninterrupted temporal coverage on the years time-scale. The best instrument up to this challenge is *Kepler*, which was launched in 7th March 2009. *Kepler* is a mission designed to detect the transit of an Earth across the Sun with its highly sensitive photometric camera; more details can be found in Basri et al. [1] and Koch et al. [2], as well as on the mission website (http://www.kepler.nasa.gov/sci).

In this chapter, I will evaluate the range of exomoons that the *Kepler Mission* or *Kepler*-class photometry (KCP) could detect through transit timing effects, with particular attention to habitable-zone exomoons. This chapter is based upon the paper Kipping et al. [3], which was co-authored with colleagues Fossey and Campanella.

I emphasise the use of *Kepler*-class photometry (KCP) throughout this work, due to the increasingly impressive results being obtained from the ground which are matching space-based photometry, for example Johnson et al. [4]. Furthermore, ground-based observations are often more ideally suited for transit-timing studies due to the fewer constraints placed on the system, such as telemetry-limited data-download speeds.

D. M. Kipping, *The Transits of Extrasolar Planets with Moons*, Springer Theses,
DOI: 10.1007/978-3-642-22269-6_7, © Springer-Verlag Berlin Heidelberg 2011

7.2 Modelling the Detectability of Exomoons

7.2.1 Confidence of Detection

In order to explore a large range of parameter space, it is more convenient and efficient to employ analytic expressions rather than repeated individual simulations for thousands of different scenarios. Therefore, one needs analytic expressions for the following:

- TTV and TDV signal amplitudes
- Times of transit minimum and transit duration errors
- Confidence of detection, based upon signal-to-noise

The TTV and TDV root mean square (RMS) amplitudes have been derived in Chap. 7, notably in (6.46) and (6.99). However, one also requires expressions for the timing errors, which are critical in evaluating the signal-to-noise.

To address this, I will use the analytic expressions for the uncertainty on the time of transit minimum (τ_T) and duration ($\tilde{T}$) as derived by Carter et al. [5] using a Fisher-analysis of a trapezoid-approximated circular-orbit light curve.

For the purposes of TDV measurements, the primary requirement is to use a measure of transit duration which has the lowest possible uncertainty. As discussed earlier in Sect. 4.2.3, by calculating the covariances of the light curve, Carter et al. [5] were able to show that $\tilde{T}$ can be calculated more precisely than either $T_{1,4}$ or $T_{2,3}$ and so I select $\tilde{T}$ as a robust duration parameter to explore the TDV effect. The uncertainties on transit depth, δ, transit duration, $\tilde{T}$, and the time of transit minimum, τ_T, were derived by Carter et al. [5] to be:

$$\sigma_\delta = W^{-1}\delta \tag{7.1}$$

$$\sigma_{\tilde{T}} = W^{-1}\sqrt{2\tilde{T}T_{1,2}} \tag{7.2}$$

$$\sigma_\tau = W^{-1}\sqrt{\tilde{T}T_{1,2}/2} \tag{7.3}$$

$$W = \delta\sqrt{\Gamma_{\mathrm{ph}}\tilde{T}} \tag{7.4}$$

where $T_{1,2}$ is the ingress/egress duration, Γ_{ph} is the photon collection rate, and δ is the transit depth. These expressions do not hold for a poorly sampled ingress or egress and therefore I assume a cadence of 1 min, corresponding to *Kepler's* short-cadence mode.

The equations of Carter et al. [5] require the ingress duration,[1] $T_{1,2}$, and the transit duration, $\tilde{T}$, as inputs, which were provided in (4.38) and (4.42) for circular orbits.

[1] Defined as the duration between contact points 1 and 2, which is generally equivalent to the egress duration.

Note, that since circular orbits will be assumed throughout this chapter, then the $T_{x,y}^{\mathrm{one}}$ expressions are equivalent to the $T_{x,y}^{\mathrm{SMO}}$ expressions. For the *Kepler Mission* or KCP, I employ the same estimate for Γ_{ph} as that of Borucki [6] and Yee and Gaudi [7]:

$$\Gamma_{\mathrm{ph}} = 6.3 \times 10^8 \mathrm{h}^{-1} 10^{-0.4(m-12)} \tag{7.5}$$

where m is the apparent magnitude. For a normal transit depth observed n times, the confidence, C, to which the transit is detected, in terms of the number of standard deviations, is defined by Borucki [6] as:

$$C(\mathrm{photometric}) = \frac{d}{\sigma_d}\sqrt{n} \tag{7.6}$$

The transit timing signals due to an exomoon are periodic in nature and so require a different detection method. Typically, this problem is tackled by searching for significant peaks in a periodogram, as often employed for radial velocity searches (e.g. Butler et al. [8]). However, this approach is less useful for transit timing effects due to an exomoon, since the frequency one is trying to detect will always be much higher than the sampling frequency, as pointed out in Sect. 6.2.4. The only available method is to therefore search for statistically significant excess variance and then use the χ^2-distribution to calculate the confidence of signal detection. In order for this method to be applicable, one requires (a) that the uncertainty estimates are robust and accurate; and (b) that the period of the signal may be derived from amplitude information alone (which may then be compared to a periodogram to further refine the frequency).

The first of these requirements can be seen to be valid as several investigations have verified. Holman et al. [9] derived the uncertainties of the times of transit minimum for four transits of XO-1b using three different methods: (i) $\Delta\chi^2 = 1$ perturbation of the best-fit; (ii) Monte Carlo bootstrapping; (iii) Markov-Chain Monte Carlo (MCMC). The authors found that all three methods produced very similar uncertainties, which implies the uncertainty estimates are highly robust. Another example is that of Carter et al. [5], who showed that the uncertainties derived using an MCMC-analysis were very similar to those predicted using analytic arguments.

The second requirement was validated in Sect. 6.5.1, where it was shown that the ratio of the TTV and TDV signal amplitudes may be used to obtain the period of the exomoon. This period may then be compared to the set of possible harmonic frequencies derived from a periodogram in order to obtain a highly reliable estimate. I therefore conclude that a search for excess variance is the most appropriate strategy for searching for exomoons through transit timing effects. The confidence of detection may be found by integrating the probability density function (PDF) of the χ^2-distribution.

$$C(\mathrm{timing}) = \sqrt{2}\mathrm{erf}^{-1}\left[1 - \int_{\alpha^2}^{\infty} \frac{x^{(n/2)-1}\exp^{-x/2}}{2^{n/2}\Gamma(n/2)}\mathrm{d}x\right] \tag{7.7}$$

where n is the number of transits observed and α^2 is the observed value of χ^2, given by:

$$\alpha^2 = n\left(1 + \frac{\delta_x^2}{\sigma_\tau^2}\right) \tag{7.8}$$

where δ_x is the RMS amplitude of the transit timing signal and σ_τ is the uncertainty on the time of transit minimum/transit duration. Integrating and making the above substitution, the confidence, C, in detecting a timing signal in units of standard deviations is:

$$C(\text{timing}) = \sqrt{2}\text{erf}^{-1}\left[1 - \text{Q}\left\{\frac{n}{2}, \frac{n}{2}\left(1 + \frac{\delta_x^2}{\sigma_\tau^2}\right)\right\}\right] \tag{7.9}$$

where Q$\{a, b\}$ is the incomplete upper regularized Gamma function. I summarise the assumptions below:

- Only one exomoon exists around the gas giant exoplanet of interest.
- The moon and planet are both on circular orbits and the moon's orbit is prograde.
- The moon's orbital plane is coaligned to that of the planet-star plane which is itself perpendicular to the line-of-sight of the observer, i.e. $i_S = i_P = \pi/2$.
- If a planet is within the habitable-zone, then any moon around that planet may also be considered to be "habitable".
- A transiting planet must be detected to 8-σ confidence to be accepted as genuine.
- An exomoon must be detected through either (a) TTV to 8-σ and TDV to 3-σ confidence or (b) TTV to 3-σ and TDV to 8-σ confidence, in order to be accepted as genuine.
- The *Kepler Mission* or KCP will be used in short cadence mode for the transit timing of a target of interest for $\simeq$4 years.
- $n = M/P_P$ where M is the mission duration and P_P is the period of transiting planet.
- At least three transits are needed to detect both a planet and a moon.

The assumption of co-alignment nullifies any TDV-TIP effect and thus the TDV signal is not enhanced by this extra contribution. In this sense, the calculations presented in this work can be considered conservative.

In most of the cases I will consider, many more than three transits will be detected and three can be seen to be the limiting case for G0V stars, where the habitable zone is sufficiently distant to only permit three transits in a 4-year timespan. Although statistically speaking three transits is sufficient, there is a risk of an outlier producing a false positive. I therefore consider detections of habitable exomoons in early G-type star systems to be described as "tentative", whereas once four transits are detected, for stars of spectral type G5V and later, this risk can be considered to be reduced.

The nominal mission length of *Kepler* is 3.5 years and it may be extended to up to 6 years, which justifies the choice of 4 years of transit timing observations. A ground-based search achieving KCP may easily be operational for 4 years or more.

I choose 8-σ as the signal detection threshold since this is the same as that used by *Kepler*. The second signal may be detected to lower significance since it is only used to confirm the phase difference between the two and also derive the exomoon period.

7.2.2 The Total Noise

The expressions of Carter et al. [5] only consider shot noise through the Γ_{ph} parameter. However, if one assumes that the impact on $\sigma_{\tilde{T}}$, σ_τ and σ_δ are approximately equivalent for additional uncorrelated noise and for correlated noise, then one may simply modify W to absorb the effects of red noise. For uncorrelated noise, I add the additional sources of noise in quadrature. Note, that this is the same treatment utilised in the design technical documents for *Kepler*, for example see Borucki [6].

In general, there are expected to be three major sources of noise present in the *Kepler* data in the form of shot noise, instrument noise and stellar variability. Instrument noise is due to a variety of effects and has been modelled in depth by Borucki [6] (and Koch [10]) to quantify its effect as a function of magnitude. With all three noise sources, I modify W to W', given by:

$$\frac{1}{W'} = \frac{1}{\delta}\sqrt{\frac{1}{\Gamma_{\mathrm{ph}}\tilde{T}} + I^2 + S^2} \tag{7.10}$$

where I is the instrument noise and S is the stellar variability. I is a function of magnitude which may be calculated using the model of Borucki [6] and I show all three noise sources plotted as a function of magnitude in Fig. 7.1. I assume a constant value for stellar variability of 10 ppm across all spectral types, a reasonable assumption, given that 65–70% of F7-K9 main-sequence stars in the *Kepler* field are likely to have similar or lower intrinsic variability than the Sun [6, 11] on timescales important to transit detections. I also note that this equation is equivalent to the formulation used in the original technical design papers for *Kepler*, for example see (7.1) of Borucki [6].

7.2.3 The Habitable-Zone

I choose to consider a moon-hosting gas-giant exoplanet around a variety of main-sequence stars as shown in Table 7.1. For each star, I calculate the habitable-zone orbital distance, a_{hab}, to be defined as the distance where a planet would receive the same insolation as the Earth. This straightforwardly permits a reasonable estimate of the habitable-zone for each star type. For a more in-depth consideration of habitability of exomoons see Williams et al. [12].

$$a_{\mathrm{hab}} = \sqrt{L_*/L_\odot}\mathrm{AU} \tag{7.11}$$

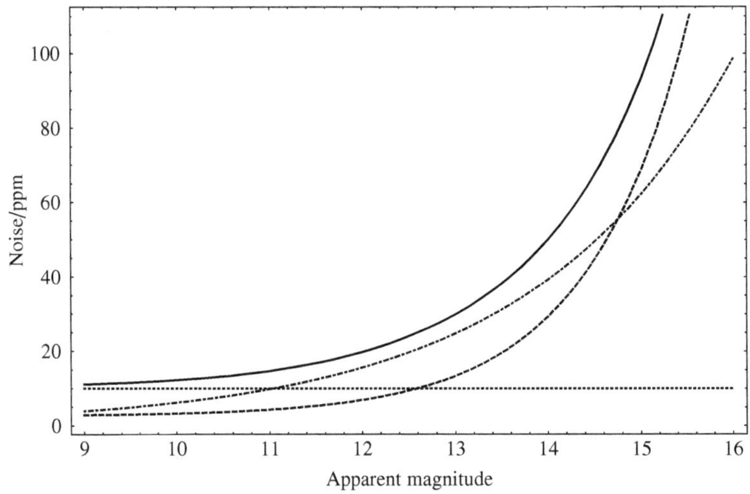

Fig. 7.1 Noise sources predicted to affect *Kepler* photometry as a function of magnitude. Instrument noise is dashed, photon noise is dot-dashed, stellar variability is dotted and the total is solid. Values obtained from Bill Borucki (personal communication)

Table 7.1 Properties of stars used in these calculations

Star type	$M_* / M_\odot$	$R_* / R_\odot$	$L_* / L_\odot$	T_{eff} /K	M_{Kep}
M5V	0.21	0.27	0.0066	3170	11.84
M2V	0.40	0.50	0.0345	3520	9.49
M0V	0.51	0.60	0.0703	3840	8.42
K5V	0.67	0.72	0.1760	4410	7.06
K0V	0.79	0.85	0.4563	5150	5.78
G5V	0.92	0.92	0.7262	5560	5.02
G2V	1.00	1.00	1.0000	5790	4.63
G0V	1.05	1.10	1.3525	5940	4.34
F5V	1.4	1.3	2.9674	6650	3.47
F0V	1.6	1.5	5.7369	7300	2.71

Values taken from Cox [13]. Absolute magnitudes in the *Kepler* bandpass calculated using guidelines on the mission website

For each planet-moon system considered, the period of the transiting planet is calculated using Kepler's Third Law:

$$P_{\mathrm{hab}} = 2\pi \sqrt{\frac{a_{\mathrm{hab}}^3}{G(M_* + M_P + M_S)}} \tag{7.12}$$

I choose to work in the time domain, rather than the orbital-distance formulation, since a major limiting factor in this study is the *Kepler Mission* duration.

7.2.4 Properties of Host Star

In this analysis, I will only consider single main-sequence stars which offer the best potential for hosting habitable environments. I will consider spectral types from M5V to F0V and assume for each an approximate mass, radius, and effective temperature as given by Cox [13], and a luminosity derived from data therein. I use the *Kepler* bandpass to calculate the absolute magnitude of these stars.[2]

For each stellar type, I assume the stars are not young and may be considered to be slow rotators. Since stellar variability is correlated to rotational period [14], I therefore limit the study to quiet stars. This is the same assumption used for *Kepler*'s ability to detect Earth-like planets since very active stars will be too variable for the detection of such bodies. In Table 7.1, I list the different star properties.

7.2.5 Properties of Exomoons

Although no exomoons have yet been discovered, it is possible to calculate the range of exomoons which are dynamically stable around each planet. Barnes and O'Brien [15] addressed this problem and developed a set of analytic expressions, which can be shown to provide excellent agreement to numerical simulations, for the stability of exomoons around exoplanets. Assuming an exomoon has to be stable for at least 5 Gyr, one is able to calculate the maximum allowed exomoon mass in each case (see (7.7–7.9) of Barnes and O'Brien [15]). I assume Jupiter-like values for the tidal dissipation factor, $Q_P = 10^5$, and for the tidal Love number, $k_{2p} = 0.51$, as used by Barnes and O'Brien [15].

One is also able to estimate the range of allowed values for the planet–moon separation, in units of Hill radii, $\mathfrak{D}$. Domingos et al. [16] presented the relevant expressions, which again can be shown to provide excellent agreement to numerical simulations. Using their (7.5), one is able to estimate the maximum distance at which an exomoon is stable for prograde orbits.

For the minimum distance, I calculate the Roche limit of the planet in all cases. If this value is greater than $2R_P$ then I use the Roche limit as the minimum distance, otherwise $2R_P$ is adopted. I assume that there is no reason for a moon to exist at any particular value of $\mathfrak{D}$ and thus the prior distribution is flat.

7.3 Light Curve Simulations

7.3.1 Light Curve Generation

A critical assumption in this paper is the use of the equations of Carter et al. [5] for the uncertainties on the transit duration and the times of transit minimum.

[2] Guidelines available from http://keplergo.arc.nasa.gov/proposal.html

The authors tested their expressions using synthetic light curves and an MCMC fitting procedure as well as a numerical Fisher-analysis. They report excellent agreement between their expressions and the derived errors but find the greatest departure for poorly sampled ingresses and near-grazing transits. The choice of assumptions made in this work avoids both of these issues.

Note that these tests were run for shot noise only and in this work I consider the effects of both instrument and stellar noise. Having modified W to W' to account for these additional noise sources, I proceed to test this modified formulation through generation of synthetic light curves.

I use the Mandel and Agol [17] code to generate light curves accounting for limb darkening based upon a non-linear law from Claret [18]. Light curves are generated for a Neptune, Saturn and Jupiter-like planet in the habitable-zone of a $m_{\mathrm{Kep}} = 12$, G2V Sun-like star with $i_P = \pi/2$ and $e_P = 0$. Each light curve is generated to have 1000 data points evenly spaced with a 1-min cadence centred on the transit minimum.

The first noise source added is shot noise generated by taking a random real number from a normal distribution of mean zero and standard deviation given by:

$$\sigma_{\mathrm{shot}} = \frac{1}{\sqrt{t_{\mathrm{exp}}\Gamma_{\mathrm{ph}}}} \tag{7.13}$$

where t_{exp} is the time between each consecutive measurement, which is assumed to be 1 min in these calculations.

For the stellar noise, 1000 sinusoidal waveforms are generated with varying periods randomly selected between 1 min up to 24 h. These waveforms are then coadded to give a single synthetic stellar signal designed to mimic the non-stochastic nature of real stellar variability. The RMS amplitude of the combined signal is then scaled to 10 ppm, which matches the amplitude prediction of Borucki [6] for a G2V star.

I consider that the instrument noise is composed of hundreds of different noise sources periodic in nature varying on timescales from 1 min to one day. There is no benefit of including timescales longer than this since transit events will not last longer than ~1 day. The combined RMS amplitude of the instrument noise is set to be given by the values calculated by Bill Borucki (personal communication) and is a function of visual magnitude. For example, a $m_{\mathrm{Kep}} = 12$ target has an RMS instrument noise of 7.47 ppm.

7.3.2 Light Curve Fitting

The noisy light curves are generated 10,000 times with the correlated noises and photon noise being randomly generated in all cases. The light curves are then passed onto a light curve fitting code used to obtain best-fit values for $\tilde{T}$ and τ_T. In all cases, I fit for $a_P/R_*, i_P, R_P/R_*$ and τ_P and therefore assume that the out-of-transit baseline is well known and its errors are essentially negligible. This is a reasonable assumption for high quality photometry with large amounts of out-of-transit data.

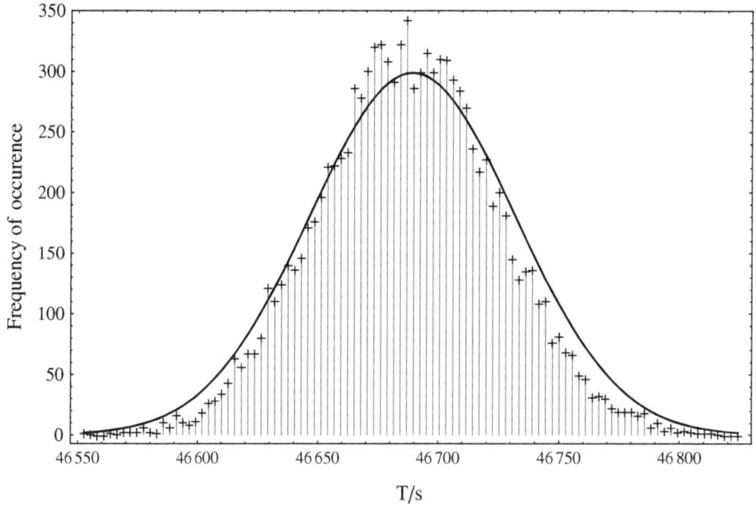

Fig. 7.2 Comparison of the distribution in $\tilde{T}$ found using Monte Carlo simulations (*scattered points*) of a transit of a Saturn-like planet versus that from theoretical prediction (*smooth line*). The theoretical prediction for the timing error (42.6 s) is slightly larger than that found in numerical simulations (37.3 s)

The fitting code finds the best-fit to the light curve by utilising a genetic algorithm,[3] PIKAIA [19], to get close to a minimum in χ^2. This approximate solution is then used as a starting point for a χ^2-minimisation performed with the AMOEBA routine [20]. The AMOEBA solution is tested by randomly perturbing it and refitting in 20 trials.

In each subsequent fitting run of the 10,000 light curves, the AMOEBA routine starts from the original best-fit. The resulting values of τ_T and $\tilde{T}$ are binned and plotted as a histogram (Fig. 7.2) and then compared to the distribution expected from the modified Carter et al. [5] expressions.

7.3.3 Comparison to the Analytic Expressions

For the three cases of a Neptune, a Saturn and a Jupiter, I obtained 10,000 estimates of $\tilde{T}$ and τ_T in each case. One may compare the distribution of τ_T and $\tilde{T}$ to the predicted distribution from the expressions of Carter et al. [5]. In all cases, I find excellent agreement between the predicted uncertainties and the theoretical values. In Fig. 7.2, I show a histogram of the results for the Saturn-case $\tilde{T}$ values, and overlay the predicted value of $\sigma_{\tilde{T}}$ for comparison where the quality of the agreement is evident from the plot. Note that this overlaid Gaussian is not a fit but a theoretical prediction.

[3] Available from http://whitedwarf.org/parallel/

I find excellent agreement for both $\sigma_{\tilde{T}}$ and σ_{τ}, and if anything the theoretical expressions slightly overestimate the uncertainties. Thus any results from this study can in fact be considered to be slightly conservative. It is preferred to adopt a conservative approach in the analysis since there may be unexpected sources of noise which increase the timing uncertainties.

7.4 Implementation of the Expressions

7.4.1 Earth-Transit Magnitude Limit

I first consider the magnitude limit in detecting an Earth-like transit which will allow one to compare the efficiency of detecting Earth-like planets and moons. *Kepler* is designed to look at $\sim 10^5$ stars between 6th to 16th magnitude.[4] For each spectral type in Table 7.1, I compute the faintest visual magnitude for which a habitable-zone Earth-like transit can be detected. *Kepler* was conceived with the goal of detecting an Earth–Sun transit of 6.5 h in duration, which represents about half the duration if the Earth has an impact parameter of zero. However, the maximum magnitude to which an Earth-like transit could be detected will be for the cases of equatorial transits as these maximise the transit duration and so too the net flux decrement. The following criteria for a reliable transit detection are defined:

- Each transit must be detected to ≥ 1-σ confidence.
- The folded light curve must have a significance of ≥ 8-σ.
- The time between contact points 2 and 3 ≥ 1 h, in order to be detected with *Kepler's* long-cadence mode of 30 min.
- At least three transits must be detected over the mission duration.

With these criteria, I compute the magnitudes of the faintest stars hosting a habitable, transiting Earth which can be detected with KCP, for both full transit durations (i.e. equatorial transits) and half-durations (i.e. impact parameter chosen to be such that the transit duration is half the full transit). The results are given in Table 7.2. As expected, smaller host stars can be fainter due to their smaller radius and hence deeper transit depths. One also picks up more transits towards smaller stars since the orbital period of the habitable-zone decreases.

Even if *Kepler* was extended to a mission length of 6 years, an F5V star would have a habitable zone so far out that detecting three transits within this region would be impossible. Hence, I do not consider F type stars in this analysis. I also consider 6 th-magnitude stars to be the bright limit, giving us d_{min}, the minimum distance for each spectral type (see Table 7.2), since the *Kepler* field is specifically chosen to avoid such bright targets.

---

[4] See http://kepler.nasa.gov/sci

Table 7.2 Faintest stars for which a habitable transiting Earth could be detected for different star types

Star type	$T_{1,4}^{\rm half} m_{\rm Kep}^{\rm max}$	$T_{1,4}^{\rm full} m_{\rm Kep}^{\rm max}$	$d_{\rm min}$ /pc	$d_{\rm max}$ /pc
M5V	–	18.111	0.68	179.56
M2V	15.992	16.190	2.00	218.78
M0V	15.206	15.465	3.28	256.45
K5V	14.278	14.618	6.14	324.79
K0V	13.286	13.712	11.07	385.83
G5V	12.762	13.240	15.70	440.56
G2V	12.236	12.763	18.79	423.25
G0V	11.520	12.106	21.48	357.44
F5V	–	–	–	–
F0V	–	–	–	–

Final column gives maximum distance of such a star with a full transit duration ($T_{1,4}^{\rm full}$), based on absolute magnitude in the *Kepler* bandpass. A blank indicates that no magnitude can satisfy the detection criteria

7.4.2 Jupiters Versus Saturns Versus Neptunes

One may use the approximate analytic expressions to get a handle on the general trends in exomoon detection. The first question one may pose is—what is the optimum planet to search for moons around, out of the three classes of Neptunes, Saturns and Jupiters? For each of these planets I simulate the detectability of the TTV signal as a function of planetary orbital period, P_P, for an 0.2 $M_\oplus$ exomoon with $\mathfrak{D} = 0.4895$. The star is fixed to a G2V spectral type with $m_{\rm Kep} = 12$ and I work with $i_P = \pi/2$ for simplicity.

I calculate the signal amplitude and time of transit minimum uncertainty in all cases and hence find the χ^2 value for a range of orbital periods. In Fig. 7.3, χ^2 is plotted as a function of host planet's orbital period. The figure reveals that Jupiters are the hardest to search for exomoons around whilst Saturns are the easiest. This is due to Saturn's low density meaning a large transit depth but a low enough mass such that an exomoon still perturbs it significantly. I find the same sequence of detectability consistently in many different orbital configurations and for TDV as well. Note that Saturn would not be considered a low-density planet from an exoplanet perspective since a significant population of much lower density exoplanets have now been found (see http://www.exoplanet.eu).

7.4.3 $\mathfrak{D}$-$m_{\rm Kep}$ Parameter Space

I now consider an exomoon in the case of a Saturn hosting a single satellite in the habitable-zone of the host star. The detectability of the timing signals depends on the planet–moon separation, $a_S = \mathfrak{D} R_{\rm Hill}$, the apparent magnitude of the host star

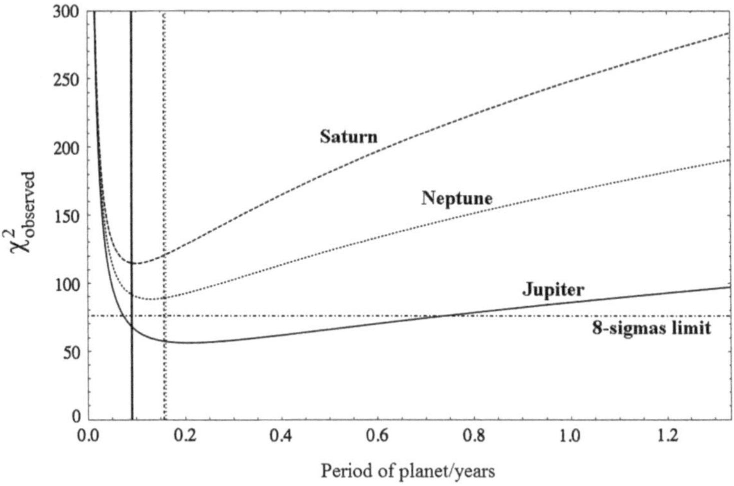

Fig. 7.3 Example plot showing the TTV detectability of a 0.2 $M_\oplus$ ($\mathfrak{D} = 0.4895$) exomoon around a Jupiter, Saturn and Neptune-like exoplanet for a G2V, $m_{\text{Kep}} = 12$ star. Values of $\chi^2 \gg 76$ are detectable at $\geq$ 8-σ confidence. Saturns present the strongest signal due to their low density. The vertical lines represent the stability limit of such a moon, calculated using the model of Barnes and O'Brien [15]

in the *Kepler* bandpass, m_{Kep}, mass of the exomoon, M_S, and finally the host star's mass and radius (i.e. spectral type).

TDV increases with lower values of $\mathfrak{D}$ and TTV increases with higher values of $\mathfrak{D}$, but one must maintain a balance such that at least one of the effects is detected to 8-σ and the other to 3-σ confidence (dictated by the previous detection criteria). To find the limits of interest, I assume a zero-impact-parameter transit as the default planet–moon arrangement. The orbital period of the host planet is always given by P_{hab}.

For a given star type and exomoon mass, I plot the contours of m_{Kep} and $\mathfrak{D}$, which provide TTVs of 3- and 8-σ confidence and then repeat for TDVs of the same confidences. There are two possible acceptance criteria: (i) TTV confidence is $\geq$3-σ and TDV confidence is $\geq$8-σ; (ii) TTV confidence is $\geq$8-σ and TDV confidence is $\geq$3-σ. The loci of points below these lines represents the potential detection parameter space. For each star type and exomoon mass, these loci will be different. In Fig. 7.4, a typical example is shown for an M0V-type star and $M_S = \frac{1}{3} M_\oplus$.

There are certain cases which constrain the stars of interest. The Barnes and O'Brien [15] limit may be calculated for a Saturn harbouring an exomoon in a system of 5 Gyr age. This suggests that the maximum moon mass that a habitable-zone Saturn could hold around a M5V star would be ~0.3 Ganymede masses, assuming the maximum prograde orbital distance is 0.4895 Hill radii, as calculated by Domingos et al. [16]. In contrast, an M2V star allows for a habitable-zone Saturn to hold onto a 0.4 Earth-mass moon for over 5 Gyr.

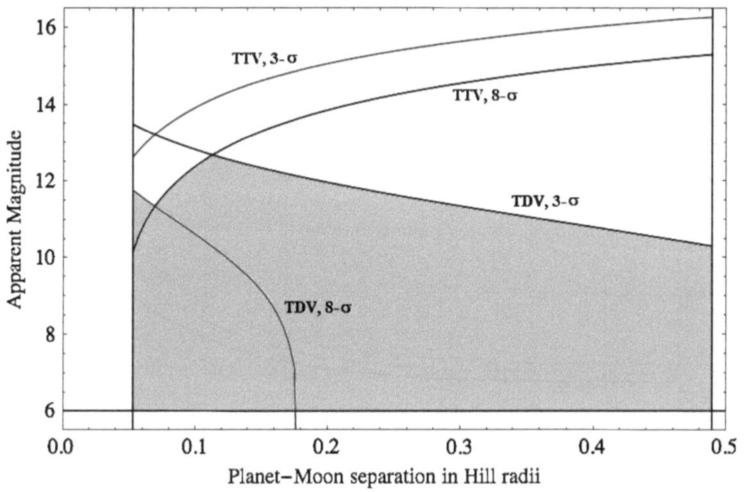

Fig. 7.4 Detectable range of a habitable $\frac{1}{3}$ $M_{\oplus}$ exomoon for an M0V star with respect to apparent magnitude and $\mathfrak{D}$ parameter space, with *Kepler*-class photometry. The *grey area* represents the area satisfying the detection criteria of detecting TTV or TDV to 8-σ and the other timing effect to 3-σ, shown by the four curves. Additional constraints are the lower Roche limit, the upper dynamical stability limit and the bright-star magnitude limit. Note how the 8-σ TDV line drops very sharply after $m \sim 9$ since *Kepler's* photometry does not improve appreciably past this magnitude due to instrument noise

A second lower limit exists from tidal forces and the Roche limit. An upper limit is given by the fact one requires three transits in a 4 year observation duration and thus the most distant habitable-zone assumed here corresponds to a period of 1.33 years, which excludes the F0V and F5V stellar types.

7.4.4 Magnitude Quartiles

For Fig. 7.4, one may convert the plot into three key numbers by assuming that the probability distribution of exomoons with respect to $\mathfrak{D}$ is approximately uniform. This simple a-priori approximation allows one to get a handle on some typical magnitude limits. This assumption effectively converts the two-dimensional array of $\mathfrak{D}_i$-$m_{\text{Kep},i}$ into a one-dimensional list of just $m_{\text{Kep},i}$. One then simply takes the quartiles of this list to obtain estimates for the 25, 50 and 75% "catch-rate" values of m_{Kep}. Physically speaking, we are saying that, for example, at the 50% catch-rate value of m_{Kep}, there is a 50% probability of detecting an exomoon if an exomoon exists with a uniform prior distribution between $\mathfrak{D}_{\text{min}}$ and $\mathfrak{D}_{\text{max}}$.

Accordingly, one may calculate the magnitude limits to detect 25, 50 or 75% of the exomoons in the given sample, which I label as the upper, middle and lower quartiles respectively. For the case shown in Fig. 7.4, the quartile values are $m_{\text{Kep}} = 12.0$, 11.5 and 10.9 respectively.

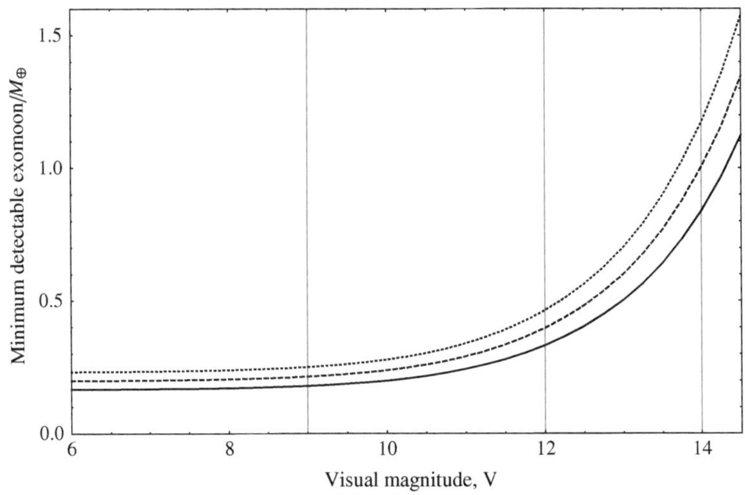

Fig. 7.5 Minimum detectable exomoon mass around a Saturn-like planet host in the habitable-zone of an M0V star. The *top line* is the 75% catch-rate, the *middle* is 50% and the *lowest* is for a 25% exomoon catch-rate. I also mark three key magnitudes of interest; 9th, 12th and 14th magnitudes

One may then calculate these quartile values for different exomoon masses, for a given star type. Effectively, one may determine the minimum detectable exomoon mass as a function of magnitude for three different detection yields: (i) 25%, (ii) 50%, and (iii) 75%. In Fig. 7.5, I show an example of one of these plots for an M2V star with a habitable-zone Saturn host planet. The contours have been calculated by a spline-interpolation of 20 evenly spaced exomoon masses. Magnitudes brighter than 6th are not considered, since *Kepler's* field-of-view has been selected to omit such stars.

7.4.5 Minimum Detectable Habitable Exomoon Masses

To complete the picture, one wishes to see how these values differ for various star types. I repeat the analysis performed above for M5V, M2V, M0V, K5V, K0V, G5V, G2V and G0V-type stars. In Fig. 7.6, I plot the minimum detectable exomoon mass, for several contours of visual magnitude, as a function of stellar mass, in the 25% detection-yield case. I also show the upper limit on moon mass calculated from Barnes and O'Brien [15] but this is only an approximation since I have assumed values of Q_P and k_{2p}.

The $m_{\mathrm{Kep}} = 12.5$ limit almost intersects the Barnes and O'Brien [15] limit at 0.4 $M_\odot$ and $\sim$0.4 $M_\oplus$ for the exomoon corresponding to an M2V star $\sim$40 pc away. For comparison, within $10,000\,\mathrm{pc}^3$ (a spherical shell of radius 13.4 pc) there are $\sim$800 main-sequence stars, $\sim$630 of which are M-dwarfs [21].

If one sets a lower limit of 10 pc for the target star, then the lowest-mass habitable-zone exomoon which could be detected with KCP around a M5V star would be

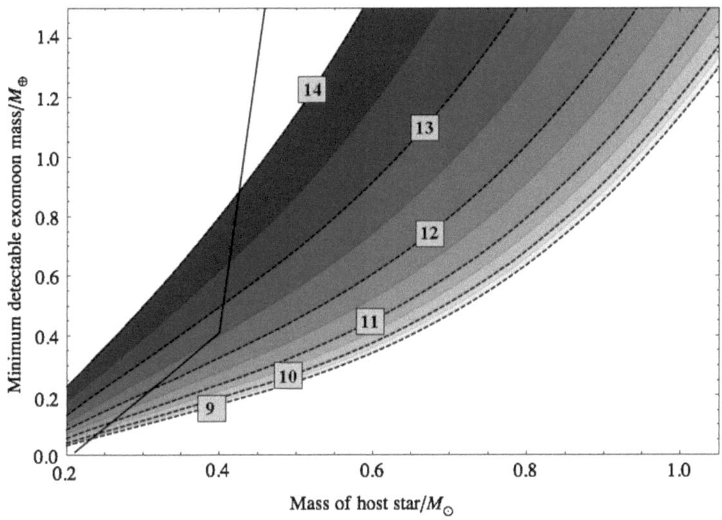

Fig. 7.6 Minimum detectable exomoon mass around a Saturn-like planet host in the habitable-zone of various star types. Contours show the different magnitude limits for a 25% exomoon-catch-rate and overlaid is the mass stability limit (*black*, *solid*), found using Barnes and O'Brien [15]

0.09 $M_\oplus$, which is above the maximum allowed stable mass limit of 0.008 $M_\oplus$. However, moving up to an M2V star I find a minimum detectable habitable exomoon mass of 0.18 $M_\oplus$ which is below the stability limit of 0.41 $M_\oplus$. I therefore conclude that the minimum detectable habitable-zone exomoon mass with KCP is ~0.2 $M_\oplus$.

7.4.6 1 $M_\oplus$ Limit

For the 25% detection yield case, one may convert the magnitude limit for detecting a 1 $M_\oplus$ habitable exomoon into a distance limit by making use of the absolute magnitudes for each star type. The distance limit for detecting a 1 $M_\oplus$ habitable exomoon may then be compared to the limit found for a transiting 1 $R_\oplus$ habitable exoplanet. In Fig. 7.7, I compare the two distance limits, which reveals that the distance limit for moons takes the same shape as the transit limit but is $\sim$ 1.6 times lower. This translates to a volume space diminished by a factor of $\sim$ 4.

It is not the intention of this work to estimate accurately how many observable stars might be contained within this volume, but one may deduce an order-of-magnitude estimate. Assuming that there are about 10^5 useable stars in *Kepler*'s field-of-view, I estimate roughly that 25,000 stars would be within range for habitable-zone exomoon detection. Extrapolating *Kepler*'s field-of-view to a band centred on the Galactic plane between $\pm$ 15°, and simplistically assuming the same number of target stars per square degree, one obtains about 2×10^6 stars that are accessible

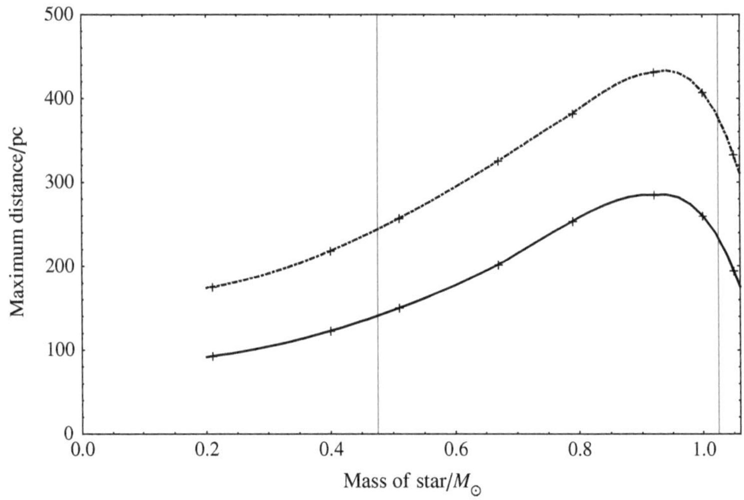

Fig. 7.7 Distance limit for detecting a 1 $M_\oplus$ habitable exomoon (*solid*) and a 1 $R_\oplus$ exoplanet (*dot-dashed*) as a function of stellar mass, for main sequence stars. I assume a Saturn-like planet as the host for the exomoon case. I also mark the lower stellar mass stability limit of $\sim 0.475\ M_\odot$ and the upper mass limit of $\sim$1 $M_\odot$ imposed by the requirement to observe $\geq$ 3 transits

to KCP for a search for habitable-zone exomoons, if each star had a habitable-zone gas giant similar to Saturn.

7.4.7 Detectability of an Earth–Moon Analogue

Finally, I consider the detectability of an Earth–Moon system. Using the orbital parameters of the Sun–Earth–Moon system and assuming the most favourable configuration of $i_P = \pi/2$, I find that the Earth would exhibit a TTV RMS amplitude of 112.4 s and a TDV of 13.7 s. The TDV is particularly low due to the large value of $\mathfrak{D}$ for the Moon.

In comparison, KCP of a $V = 6$ star would yield timing errors of $\sigma_{t_c} = 389.6$ s and $\sigma_T = 779.2$ s. Even for the most favourable case then, the TTV has a signal-to-noise of 0.3, which would be quite undetectable. I therefore conclude that KCP cannot detect a Sun–Earth–Moon analogue through transit timing effects.

7.5 Conclusions

In this chapter, I have presented a feasibility study which finds that habitable exomoons are detectable up to $\sim$100–200 pc away around early-M, K and later-G dwarf stars with the *Kepler Mission* or photometry of equal quality (KCP). This photometric quality should be sensitive down to 0.2 $M_\oplus$ habitable exomoons in the idealised cases and could survey $\sim 10^6$ stars for 1 $M_\oplus$ habitable exomoons, with a full Galactic Plane survey. This number is around 25,000 stars for *Kepler*'s field-of-view.

Saturn-like planets are found to be the ideal host candidates for detection, due to their large radius to mass ratio. Additionally, I find lower-mass exomoons may be found around M-dwarfs due to the closer habitable-zone permitting a larger number of transits in a 4-year window.

The exciting prospect of discovering a habitable exomoon is well within the grasp of KCP and whether such worlds exist or whether they will be classed as truly habitable worlds are questions we can hope to answer in the coming years. It is interesting to note that $\sim 0.2\,M_\oplus$ is the minimum habitable mass for a planet or moon used by several other sources in the literature, including Raymond et al. [22] ($0.3\,M_\oplus$) and Williams et al. [12] ($0.12\,M_\oplus$). It has also been proposed by Scharf [23] that the habitable-zone could be extended for exomoons due to tidal heating and so these calculations may in fact be an underestimate.

I find that *Kepler* will be incapable of finding Ganymede-mass moons around Saturn-like planets. Canup and Ward [24] have suggested that the maximum moon mass which could form from a planetary debris disk is 2×10^{-4} of that of the primary's and therefore, if this rule holds true, it unlikely KCP would detect any moons which formed around a planet in such a way. However, moons above this mass limit are still dynamically stable (according to Barnes and O'Brien [15]) and could have been captured by a planet or formed through an impact, for example like the formation of Triton and the Moon respectively (see Chap. 2 for a more detailed discussion). Whether or not such objects are common is unknown but KCP could make the first in-depth search.

These results suggest it is easier to detect an Earth-like exoplanet than an Earth-like exomoon around a gas giant. However, there are no statistics to draw upon to estimate which of these scenarios is more common. If a roughly equal number of both are discovered, it would indicate that the latter is more common due to the detection bias.

These results highlight the promising opportunity of making the first exomoon detection using the *Kepler* telescope, or photometry of equivalent quality, especially the feasibility of detecting habitable-zone exomoons. All-sky surveys focussing on bright M-dwarf stars would be ideally placed to search for habitable exomoons in greater depth and thus a telescope like *TESS* or *PLATO* could continue the search after the *Kepler Mission* ends.

References

1. G. Basri, W.J. Borucki, D. Koch, The Kepler Mission: a wide-field transit search for terrestrial planets [review article]. New Astron. Rev. **49**, 478–485 (2005)
2. D. Koch, W. Borucki, G. Basri, T. Brown, D. Caldwell, J. Christensen-Dalsgaard, W. Cochran, E. Devore, E. Dunham, T.N. Gautier, J. Geary, R. Gilliland, A. Gould, J. Jenkins, Y. Kondo, D. Latham, J. Lissauer, D. Monet, The Kepler Mission and eclipsing binaries, in: *IAU Symposium, vol. 240 of IAU Symposium*, 2007, ed. by W.I. Hartkopf, E.F. Guinan, P. Harmanec, pp. 236–243. doi:10.1017/S1743921307004085

3. D.M. Kipping, S.J. Fossey, G. Campanella, On the detectability of habitable exomoons with Kepler-class photometry. Mon. Not. R. Astron. Soc. **400**, 398–405 (2009)
4. J.A. Johnson, J.N. Winn, N.E. Cabrera, J.A Carter, A smaller radius for the transiting exoplanet WASP-10b. Astrophys. J. Lett. **692**, L100–L104 (2009)
5. J.A. Carter, J.C. Yee, J. Eastman, B.S. Gaudi, J.N. Winn, Analytic approximations for transit light-curve observables, uncertainties, and covariances. Astrophys. J. **689**, 499–512 (2008)
6. W.J. Borucki et al., Kepler Mission: design, expected science results, opportunities to participate, *A Decade of Extrasolar Planets around Normal Stars*, vol. 1 of Cambridge, 2005, ed. by M. Livio, K. Sahu, J. Valenti, pp. 36–+
7. J.C. Yee, B.S. Gaudi, Characterizing long-period transiting planets observed by Kepler. Astrophys. J. **688**, 616–627 (2008)
8. R.P. Butler, G.W. Marcy, S.S. Vogt, C.G. Tinney, H.R.A. Jones, C. McCarthy, A.J. Penny, K. Apps, B.D. Carter, On the double-planet system around HD 83443. Astrophys. J. **578**, 565–572 (2002)
9. M.J. Holman, J.N. Winn, D.W. Latham, F.T. O'Donovan, D. Charbonneau, G.A. Bakos, G.A. Esquerdo, C. Hergenrother, M.E. Everett, A. Pál, The transit light curve project. I. Four consecutive transits of the exoplanet XO-1b. Astrophys. J. **652**, 1715–1723 (2006)
10. D. Koch, A model for estimating the number of stars for which terrestrial planets can be detected using transits, *Bioastronomy 2002: Life Among the Stars, volume 213 of IAU Symposium*, June 2004, ed. by R. Norris, F. Stootman, pp. 85–+
11. N.M. Batalha, J. Jenkins, G.S. Basri, W.J. Borucki, D.G. Koch, Stellar variability and its implications for photometric planet detection with Kepler, *Stellar Structure and Habitable Planet Finding, volume 485 of ESA Special Publication*, January 2002, ed. by B. Battrick, F. Favata, I.W. Roxburgh, D. Galadi, pp. 35–40
12. D.M. Williams, J.F. Kasting, R.A. Wade, Habitable moons around extrasolar giant planets. Nature **385**, 234–236 (1997)
13. A.N. Cox, Allen's astrophysical quantities, 4th edn, ed. by A.C. Cox (Springer AIP Press, NewYork, 2000) ISBN:0-387-98746-0
14. J.D. Dorren, E.F. Guinan, L.E. Dewarf, The decline of solar magnetic activity with age, *Cool Stars, Stellar Systems, and the Sun, volume 64 of Astronomical Society of the Pacific Conference Series*, 1994, ed. by J.-P. Caillault, pp. 399–+
15. J.W. Barnes, D.P. O'Brien, Stability of satellites around close-in extrasolar giant planets. Astrophys. J. **575**, 1087–1093 (2002)
16. R.C. Domingos, O.C. Winter, T. Yokoyama, Stable satellites around extrasolar giant planets. MNRAS **373**, 1227–1234 (2006)
17. K. Mandel, E. Agol, Analytic light curves for planetary transit searches. Astrophys. J. Lett. **580**, L171–L175 (2002)
18. A. Claret, A new non-linear limb-darkening law for LTE stellar atmosphere models. Calculations for $-5.0 <= \log[M/H] <= +1$, 2000 K <= Teff <= 50000 K at several surface gravities. Astron. Astrophys. **363**, 1081–1190 (2000)
19. T.S. Metcalfe, P. Charbonneau (2003) Stellar structure modeling using a parallel genetic algorithm for objective global optimization. J. Comput. Phys. **185**, 176–193 (2003). doi:10.1016/S0021-9991(02)00053-0
20. W.H. Press, S.A. Teukolsky, W.T. Vetterling, B.P. Flannery, *Numerical recipes in FORTRAN. The art of scientific computing* (Cambridge University Press, Cambridge, 1992)
21. G. Ledrew, The real starry sky. J. R. Astron. Soc. Can. **95**, 32–+ (2001)
22. S.N. Raymond, J. Scalo, V.S. Meadows, A decreased probability of habitable planet formation around low-mass stars. Astrophys. J. **669**, 606–614 (2007)
23. C.A. Scharf, The potential for tidally heated icy and temperate moons around exoplanets. Astrophys. J. **648**, 1196–1205 (2006)
24. R.M. Canup, W.R. Ward, A common mass scaling for satellite systems of gaseous planets. Nature **441**, 834–839 (2006)

Chapter 8
Conclusions and Future Work

For tomorrow belongs to the people who prepare for it today
African Proverb

8.1 Summary

Twenty years ago, humanity knew of just a handful of planets, most of which had been known since antiquity. Since this time, a new branch of astronomy has blossomed and bore many fruits. At the time of writing, over 500 exoplanets are known to exist and the breadth and depth of understanding of those worlds has matured from measuring just minimum masses to characterizing their atmospheres, dynamics and temperatures. The rich history of planet and exoplanet detection is overviewed in Chap. 1.

The ultimate goal of many exoplanetary scientists is to detect an Earth-like world; a habitable, temperate abode where beings such as ourselves may just reside and ask the same questions humanity has asked since the dawn of civilization. Only now can such worlds be detected and early hints suggest Earths may be frequent [1] and a detection is round the corner [2].

The search for an Earth-like world is centered almost exclusively upon the search for an Earth-like planet, but this may be unnecessarily restrictive—Earth-like moons may also be common throughout the Universe. A moon in the habitable-zone with a mass $\geq 0.3\,M_{\oplus}$ could retain an atmosphere and satisfy all of the criteria usually associated with habitable planets [3]. In Chap. 2, I reviewed the likely properties of extrasolar moons based upon physical constraints. The conclusion was that for a moon to be $\geq 0.3\,M_{\oplus}$, a capture or impact origin is probably necessary. With only two known examples of large satellites originating in such a way, the frequency of habitable moons is unknown and a Copernican view does not favour such a hypothesis. However, it must also be acknowledged that a Copernican model of planetary systems is an extremely poor one, with most exoplanetary systems bearing little resemblance to our own.

D. M. Kipping, *The Transits of Extrasolar Planets with Moons*, Springer Theses, 183
DOI: 10.1007/978-3-642-22269-6_8, © Springer-Verlag Berlin Heidelberg 2011

With every reason to conduct a search for such bodies, the missing piece in the puzzle is a method to actually detect moons. Whilst several methods have been proposed for non-transiting systems, which are reviewed in Chap. 2, it is the transiting system which has always been the most generous donor of information for distant worlds. This forms the principal goal of the thesis presented here—a method to detect exomoons in transiting systems.

Before one can search for the tiny perturbations due to a moon buried within the planet's transit signal, one must first describe a detailed model of transiting planets by themselves. In Chap. 3, I presented the basics of transiting planets, reviewing the current literature on the subject. This chapter introduces the various assumptions made in the model too, and discusses the consequences of breaking each one.

It was shown later that exomoons can be detected through changes in the transit times and durations. Therefore, a necessary step is a detailed understanding of the transit duration too. Chapter 4 is dedicated to the times of transit minima and the transit duration. It is shown that in both cases an exact, simple, closed-form expression is elusive and one is forced to deal with solutions from a bi-quartic equation. Due to the root selection issues introduced by the bi-quartic, it is preferable to adopt alternative approaches.

For the transit minima, I have presented a series expansion solution up to 6th-order in $\cos^2 i_P$ which rapidly converges upon the true minima solution (Eqs. (4.18) and (4.25) for the transit and occultation minima respectively). This series was previously known to the second-order, but the additional terms allow one to achieve much higher precisions which are frequently required in modern exoplanet timing.

For the transit duration, a physically motivated new approximate formula is derived which is shown to yield the highest accuracy out of all previously proposed formulas (4.54). The equation is easily differentiated allowing one to evaluate the effects of secular parameter change, such as apsidal precession, nodal precession, etc. Eqs. (4.86–4.91).

In Chap. 3, three assumptions in the transit model were identified as being frequently invalid and to have significant consequences in such a scenario. These are A0) a uniform source star; A4/A5) the planet emits no flux/there are no background luminous objects; A10) the integration times are small. A0 can be compensated for by using the Mandel and Agol [4] code to account for limb darkening and generally avoid observing spotty stars. In Chap. 5, I addressed the consequences of methods of compensation for assumptions A4/5 and A10.

The first of these is the case of a blended source of light in addition to the host star. The blend source could be a background object or even the planet itself for very hot planets observed at infrared wavelengths. In all cases, the blend leads to a dilution of the transit depth. In Sect. 5.2, I presented a model for compensating for blend sources, with particular focus on the self-blend scenario of a hot-planet, since this was previously unconsidered.

For the other assumption, A10, it is shown in Sect. 5.3 that long-integration times lead to a smearing of the transit light curve signal. This smearing severely compromises the parameter retrieval, unless accounted for. In the example of the 30-min integrations of the *Kepler Mission*, the effects are so severe that many planets would be rejected as unphysical unless the integration time is included in the model. Sect. 5.3 describes the appropriate model to accomplish this and how to quickly evaluate the required numerical integration needed.

With the transit model now well-understood, one may return to the original problem of this thesis—how to detect exomoons in transiting systems. In Chap. 6, I introduced the transit timing variations (TTV) method first proposed by Sartoretti and Schneider [5]. The premise is that the moon causes the planet to exhibit reflex motion on top of the orbital motion around the star leading to variations in the position of the planet. These positional changes manifest as offsets to the times of transit minimum. It is shown that the TTV signal is always heavily undersampled due to the orbital constraints of the moon, which means one cannot determine a period for the exomoon. Without this information, TTV only provides mass of the moon multiplied by the distance to the planet, $a_S M_S$. This was identified as one of two critical problems with TTV in Sect. 6.2.4. The other problem was the plethora of other effects which also produce TTV, meaning that one requires a method to discriminate between an exomoon TTV and the other sources.

Before tackling these problems, Sect. 6.3 provides an updated model for the TTV accounting for many effects not given in the Sartoretti and Schneider [5] model, such as orbital eccentricities and non-coplanarity. I also provide a very careful and detailed outline of the various assumptions required in the model. Next, the prediction of velocity induced transit duration variations (TDV-V) is presented in Sect. 6.4. The planet's reflex motion results in changes in not only position but also velocity and this causes a broadening and shrinking of the transit duration.

It is shown that TDV-V solves both of the problems with TTV alone. Firstly, TDV-V scales as M_S/P_S meaning the ratio of the TDV-V to TTV RMS amplitudes (which is determined through excess variance analysis) provides P_S and M_S separately. Second, TDV-V exhibits a $\pi/2$ phase shift from TTV which gives the observer a unique exomoon signature to hunt for.

In Sect. 6.6, a second-order effect is explored which I denote as transit impact parameter induced transit duration variations (TDV-TIP). Here, the moon causes the planet to exhibit a component of its reflex motion orthogonal to the planet's trajectory across the stellar face. These orthogonal kicks cause the transit impact parameter to vary, which the duration is very sensitive to. The effect is maximized for highly inclined moons, but such systems are expected to be rare due to the contracted region of Hill stability [6]. For co-aligned moons, the effect is still present but generally an order-of-magnitude less than the TDV-V effect. Still, high signal-to-noise measurements may be able to exploit the effect to determine the sense of orbital motion. This is because for prograde orbits TDV-V and TDV-TIP are in phase and thus constructively interfere, but for retrograde orbits the phase difference is π causing destructive interference. I direct the reader to Table 6.1 for a concise summary of the properties of the various timing effects.

The TTV and TDV effects therefore offer a way of detecting extrasolar moons in transiting systems, satisfying the original goal of this thesis. However, until this point, it remains unclear how feasible such an enterprise actually is. Therefore, in Chap. 7, I present a feasibility study for detecting exomoons with the *Kepler Mission*,

Fig. 8.1 Artist's impression of a habitable exomoon. Kepler may well detect such an object in the near future. *Credit* Dan Durda

or photometry of equal quality. The study focusses exclusively on habitable-zone exomoons, since this not only limits the parameter space of investigation, but also is the case of greatest interest (Fig. 8.1).

The results find that habitable-zone exomoons down to 0.2 $M_{\oplus}$ may be detectable with *Kepler*. The ideal host planets are low-density planets due to the large transit signal but high reflex motion e.g. Saturn. The ideal host stars are small and cool to give a large transit depth and short period habitable-zone e.g. an M-dwarf. The analysis finds that $\sim$25,000 stars may be surveyed within *Kepler's* field-of-view for 1 $M_{\oplus}$ habitable-zone exomoons. This exciting prospect means that *Kepler* will be able to obtain meaningful constraints on the frequency of habitable-zone exomoons in the coming years. It also validates TTV and TDV as detection techniques, showing that these methods are not just theoretical oddities but plausible, viable detection techniques.

One of the limitations of *Kepler* is that the telescope is a deep-field survey meaning most of the target stars are faint by the standards of radial velocity surveys. Since the transit method cannot provide the mass of a planet, only its radius, then if and when *Kepler* detects an Earth-sized planet there will be no way of knowing what the mass of the planet is. As a result, the composition will be completely unknown which is clearly key to understanding the likely habitability. In fact, without radial velocity confirmation the planetary nature of the signal is also questionable with numerous astrophysical false positives being equally viable. It is in this respect that exomoons will have a potent advantage.

Extrasolar moons can have their masses determined through the TTV and TDV effects. The only requirement is a prior knowledge of the planetary mass through radial velocity measurements, but given that the planetary host may well be a Jupiter-sized object, RV confirmation will be trivial relative to the Earth-mass case. Indeed, recently Kipping [7] has gone further by showing that the exomoon systems could even be used to weigh the mass of the star directly which removes any stellar evolution dependencies which are usually invoked. The TTV/TDV signals may be used to compile an exomoon ephemeris and predict the moments when the moon transits.

This would allow for the radius of the moon to be measured as well and thus the mass and radius would be known. Finally then, one can see that the composition and structure of the moon could be inferred by combining the moon's mass and radius with structural models (e.g. Valencia et al. [8]). In this sense then, not only would the exomoon detection be more secure, it would also provide a great deal more information about the nature of the object.

8.2 Future Work

For exolunar science, there is a great deal of work to be done in both the theoretical and observational aspects. The observational work is fairly obvious since no exomoons have been detected yet; a dedicated program to search for their existence is required. I will here though consider the theory side, which is most relevant to the theme of this thesis. The model for TTV and TDV presented in this thesis can be considered as really the first step in the long chain of future developments which can be envisaged and ultimately required. I list here some aspects which I identify as requiring further work for the TTV and TDV theory specifically:

- Generalization of TTV and TDV to multiple moons
- Generalization to multiple planet systems, which can also induce TTV/TDV
- Extension of TTV and TDV to more general three-body cases, including long-term variations e.g. nodal precession
- Solutions for the exact integrals of the TTV and TDV RMS amplitudes using sky-projected distances rather than just the dominant Cartesian component.

One major problem which remains unsolved is the modelling of the transit light curve of a planet with a moon in a completely analytic manner. This model needs to include mutual events, eccentricity, the full three-dimensional orbital elements, stellar limb darkening and inherently account for TTV and TDV. This demanding list of requirements is challenging enough, but the entire model must be completely analytic for a light curve fitting code to have any chance of expediently fitting real observations. Whilst numerical techniques such as pixelating the stellar disc are much easier to write down, the analytic solution is the real prize offering the potential for rapid exploration of the large numbers of model parameters.

Although not a part of this thesis, I mention here a work in preparation which has indeed solved this problem. The framework and code for a completely analytic planet–moon transit modelling routine, called `LUNA`, has been developed by myself in recent months. `LUNA` is a major step forward as it will not only model the TTVs and TDVs, but also previously unaccounted for effects such as velocity induced ingress/egress asymmetry. Further, `LUNA` models the exomoon's light curve dip and thus the moon's own reflex motion (including the moon's own version of TTV, TDV-V, TDV-TIP, etc.) is automatically accounted for. Finally, `LUNA` also allows one to measure the radius of the moon through the depth of the transit features it imparts. Put simply, when implemented in a fitting routine, `LUNA` will extract every ounce

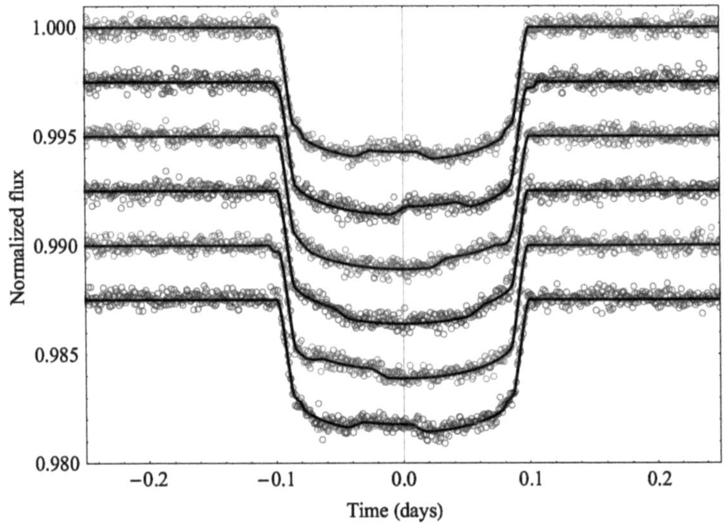

Fig. 8.2 Example light curve from the LUNA code for a moon at 5% of the Hill radius around a habitable-zone Neptune around an M2V star. Normal TTV/TDV methods are invalid here due to the moon's very short period but the moon can be easily detected through the crossing events as it eclipses the planet during the transits. This is a work in preparation

of information imparted onto the light curve by the moon. It will be a highly potent weapon in exomoon searches (for example see Fig. 8.2 and 8.3).

Despite the obvious power of LUNA, TTV and TDV will remain as very useful methods for finding exomoons. This is because LUNA is a moon-specific algorithm whereas TTV and TDV can be obtained by just modelling a planetary transit. It is now becoming a standard aspect of new observations of transiting systems to provide the transit timings allowing one to use archived timings to search for exomoons, perturbing planets or whatever else an observer is interested in. In contrast, LUNA requires the full photometric time series and it is clearly a more specialized and dedicated algorithm.

LUNA is currently just in the beta-phase and models the transits for a planet with a single moon, so another obvious extension is to include multiple moons. For the time being, a more paramount goal is probably to implement LUNA on *Kepler* photometry, which I will be doing in the future.

This then is a time to be hopeful for finding habitable worlds. An Earth-like exoplanet could be announced any day now, and the methods for finding Earth-like moons are now becoming mature enough for effective implementation in a survey style. Then it seems great insight will be soon forthcoming regarding the question which has haunted great thinkers for millennia—*are we alone?*

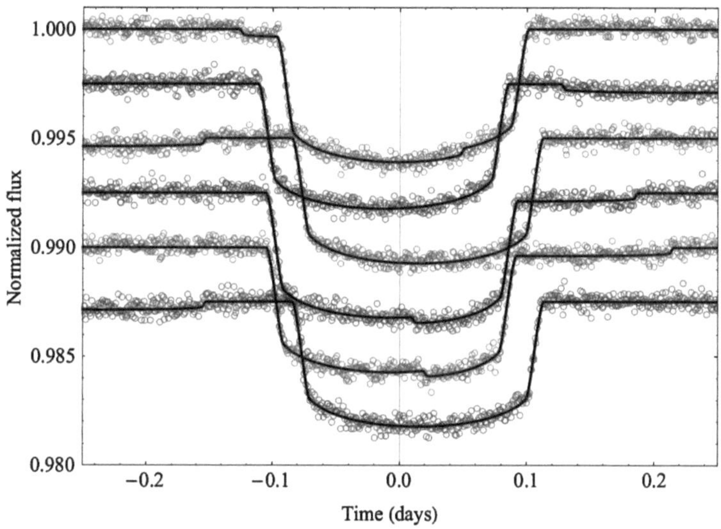

Fig. 8.3 Example light curve from the LUNA code for a moon at the Hill radius around a habitable-zone Neptune around an M2V star. The TTV effect is visible by eye but crossing events are less probable due to the moon's much longer period. This is a work in preparation

References

1. A.W. Howard, G.W. Marcy, J.A. Johnson, D.A. Fischer, J.T. Wright, H. Isaacson, J.A. Valenti, J. Anderson, D.N.C. Lin, S. Ida, The occurrence and mass distribution of close-in Super-Earths, Neptunes, and Jupiters. Science **330**, 653–655 (2010)
2. S. Arbesman, G. Laughlin, A scientometric prediction of the discovery of the first potentially habitable planet with a mass similar to Earth. PLoS ONE **5**, e13061 (2010). doi:10.1371/journal.-pone.0013061
3. D.M. Williams, J.F. Kasting, R.A. Wade, Habitable moons around extrasolar giant planets. Nature **385**, 234–236 (1997)
4. K. Mandel, E. Agol, Analytic light curves for planetary transit searches. Astrophys. J. Lett. **580**, L171–L175 (2002)
5. P. Sartoretti, J. Schneider, On the detection of satellites of extrasolar planets with the method of transits. Astron. Astrophys. Suppl. **134**, 553–560 (1999)
6. J.R. Donnison, The Hill stability of inclined small mass binary systems in three-body systems with special application to triple star systems, extrasolar planetary systems and Binary Kuiper Belt systems. Planet. Space Sci. **58**, 1169–1179 (2010)
7. D.M. Kipping, How to weigh a star using a moon. Mon. Not. R. Astron. Soc. **409**, L119–L123 (2010). doi:10.1111/j.1745-3933.2010.00961.x
8. D. Valencia, R.J. O'Connell, D. Sasselov, Internal structure of massive terrestrial planets. Icarus **181**, 545–554 (2006)

Appendix A
Notations and Acronyms

A.1 Discussion of Notation Used

A.1.1 Introduction

During the writing of this thesis, a great effort was spent in finding a self-consistent notation spanning the full breadth of transiting planets and moons. Within the field of transiting exoplanets, a wide berth of notation and symbolic representation is employed. Some of the choices are historic, some convenient, and others enigmatic. The goal of the scheme I used, was to be able to convey parameters unambiguously, even if that comes at the cost of a somewhat more elaborate notation scheme.

A.1.2 Remaining Ambiguities

Despite my efforts, some ambiguities do still exist. For example, early on in the thesis, whilst discussing a planet orbiting a star only, the planetary properties, such as the period and semi-major axis, are given the subscript "P" to denote planet. However, once a planet-moon system is introduced, the period of the planet around the star becomes an ambiguous definition and really one now has two periods: P_B the period of the planet-moon barycentre around the star and P_S, the period of the satellite around the planet-moon barycentre (which must equal the period of the planet around the planet-moon barycentre). Whilst P_B is a more encompassing definition, it would be confusing to employ P_B rather than P_P early on in the thesis as the barycentre would be in reference to the barycentre of the planet itself, which is implicit with the standard definition of P_P anyway. Therefore, to remain consistent, P_P is used for the orbital period of the planet-moon barycentre around the star later on too.

D. M. Kipping, *The Transits of Extrasolar Planets with Moons*, Springer Theses,
DOI: 10.1007/978-3-642-22269-6, © Springer-Verlag Berlin Heidelberg 2011

A.1.3 Subscript Scheme

Subscripts to various parameters come in three flavours. The first is the simple case of a single subscript letter e.g. P_P. In this case, the subscript letter refers to a type of object e.g. "P" for planet, "S" for satellite, "*" for star, "b" for planet b, "c" for planet c, etc.

The second type of subscript used is two letters without any separation e.g. S_{P*}, X_{S*}, etc. Both letters again refer to objects but here the term in general means object A to object B. For example, S is the sky-projected separation between two objects, so for S_{P*} one is referring to the sky-projected planet-star separation.

The third type of subscript used is with a comma separating the two letters e.g. $b_{P,T}$, $\varrho_{P,O}$, etc. This refers to the parameter in reference to the object as the first letter but evaluated at or in reference to some special instance denoted by the second letter. For example, $f_{P,T}$ refers to the true anomaly of the planet, "P", evaluated at the time of transit minimum, which is always given by the subscript "T". The other instance commonly used is "O" for occultation.

A.1.4 Superscripts

In some cases, superscripts are used to further the definition of a parameter. This is generally done because otherwise the subscripts would become very messy whereas the superscripts remain an empty slot. In this thesis, superscripts generally denote some additional property of a parameter and there is a wide range of its usage. For example, O_{xy}^{one} denote that O_{xy} is calculated using the "one"-term model for the transit duration. Other options include the "two"-term model and references to different authors. Another example would be ρ_*^{circ}, which refers to the density of the star evaluated for circular orbits ("circ").

A.2 Nomenclature

There are also various naming schemes employed throughout this thesis which I believe require some explanation.

A.2.1 "Satellite" vs "Moon"

Throughout this thesis, the companions of extrasolar planets are referred to as "exomoons" or "moons" rather than "exosatellites" or "satellites". Whilst a satellite can denote anything in orbit of a planet, a moon refers specifically to a natural satellite and therefore is the more apt description for what I am searching for.

Despite this, the subscript "S" is always used in reference to satellite. The reason for this choice is that "S" is the same terminology used by numerous other authors in regard to exomoons, starting with Sartoretti and Schneider (1999) and consistently used since. This is an example of a historically consistent nomenclature.

A.2.2 "Occultation" vs "Secondary Eclipse"

In the exoplanet literature, it is generally more common to denote the anti-transit as the secondary eclipse rather than the occultation. However, in this thesis, occultation is always used. If one uses secondary eclipse, then the transit is dubbed the primary transit. Therefore, a natural choice of subscripts would be "P" for primary and "S" for secondary. However, both "P" and "S" subscripts are already used in reference to the planet and satellite. Therefore, this is a not a viable choice for our purposes.

Instead, "transit" and "occultation" give the subscripts "T" and "O" which are unambiguous against other subscript terms.

A.2.3 "Transit Minimum" vs "Mid-Transit Time"

In the exoplanet literature, the timing of the transit is usually given by what is called the "mid-transit time". Throughout this thesis, the nomenclature of "mid-transit time" is carefully avoided and "transit minimum" or time of the transit minimum is used throughout. This is because the mid-transit time is highly ambiguous. For an eccentric orbit, the halfway point (i.e. the mid-point) between the first and last contacts is not the same as the instant of inferior conjunction. Further, the instant of inferior conjunction is not the same as the instant when the sky-projected planet-star separation is minimized. For a limb-darkened star, the minimum of the transit curvature occurs when S_{P*} is minimized and this represents a completely unambiguous definition.

For this reason, "transit minimum" is always used. This invalidates several of the standard terms used to reference this point in time e.g. t_C referring to central time and t_{mid} referring to mid-time. In light of the ambiguity of mid-times, these terms can be seen to be unsatisfactory descriptors of the instant of the transit minimum. For that reason, a new notation is devised, τ, which alleviates these problems.

A.3 List of Acronyms

Table A.1 provides a list of the various acronyms used throughout this thesis.

Table A.1 List of important acronyms used throughout this thesis

Acronym	Definition
TTV	Transit timing variations
TDV	Transit duration variations
TDV-V	Velocity induced transit duration variations
TDV-TIP	Transit impact parameter induced transit duration variations
MCMC	Markov chain Monte Carlo
RMS	Root-mean-square
SNR	Signal-to-noise
FAP	False alarm probability
RV	Radial velocity

A.4 List of Parameters

Table A.2 provides a list of the various key parameters used throughout this thesis.

Table A.2 List of important parameters used in this paper

Parameter	Definition
S'	Sky-projected separation between two objects
S	Sky-projected separation between two objects, in units of R_*
X'	Sky-projected X-component position of an object
X	Sky-projected X-component position of an object, in units of R_*
Y'	Sky-projected Y-component position of an object
Y	Sky-projected Y-component position of an object, in units of R_*
Z'	Sky-projected Z-component position of an object
Z	Sky-projected Z-component position of an object, in units of R_*
S'_{P*}	Sky-projected planet-star separation
S_{P*}	Sky-projected planet-star separation, in units of R_*
X'_{P*}	Sky-projected X-component of planet-star separation
X_{P*}	Sky-projected X-component of planet-star separation, in units of R_*
Y'_{P*}	Sky-projected Y-component of planet-star separation
Y_{P*}	Sky-projected Y-component of planet-star separation, in units of R_*
Z'_{P*}	Sky-projected Z-component of planet-star separation
Z_{P*}	Sky-projected Z-component of planet-star separation, in units of R_*
S'_{S*}	Sky-projected satellite-star separation
S_{S*}	Sky-projected satellite-star separation, in units of R_*

(continued)

Table A.2 (continued)

Parameter	Definition
X'_{S*}	Sky-projected X-component of satellite-star separation
X_{S*}	Sky-projected X-component of satellite-star separation, in units of R_*
Y'_{S*}	Sky-projected Y-component of satellite-star separation
Y_{S*}	Sky-projected Y-component of satellite-star separation, in units of R_*
Z'_{S*}	Sky-projected Z-component of satellite-star separation
Z_{S*}	Sky-projected Z-component of satellite-star separation, in units of R_*
S'_{S*}	Sky-projected satellite-star separation
S_{S*}	Sky-projected satellite-star separation, in units of R_*
X'_{S*}	Sky-projected X-component of satellite-star separation
X_{S*}	Sky-projected X-component of satellite-star separation, in units of R_*
Y'_{S*}	Sky-projected Y-component of satellite-star separation
Y_{S*}	Sky-projected Y-component of satellite-star separation, in units of R_*
Z'_{S*}	Sky-projected Z-component of satellite-star separation
Z_{S*}	Sky-projected Z-component of satellite-star separation, in units of R_*
S'_{PS}	Sky-projected planet-satellite separation
S_{PS}	Sky-projected planet-satellite separation, in units of R_*
X'_{PS}	Sky-projected X-component of planet-satellite separation
X_{PS}	Sky-projected X-component of planet-satellite separation, in units of R_*
Y'_{PS}	Sky-projected Y-component of planet-satellite separation
Y_{PS}	Sky-projected Y-component of planet-satellite separation, in units of R_*
Z'_{PS}	Sky-projected Z-component of planet-satellite separation
Z_{PS}	Sky-projected Z-component of planet-satellite separation, in units of R_*
f	True anomaly of an object around its primary
E	Eccentric anomaly of an object around its primary
M	Mean anomaly of an object around its primary
f_P	True anomaly of the barycentre of the planet (+any satellites) around the host star
E_P	Eccentric anomaly of the barycentre of the planet (+any satellites) around the host star
M_P	Mean anomaly of the barycentre of the planet (+any satellites) around the host star
f_S	True anomaly of the satellite around the host planet
E_S	Eccentric anomaly of the satellite around the host planet
M_S	Mean anomaly of the satellite around the host planet
e	Orbital eccentricity of an object around its primary
e_P	Orbital eccentricity of the barycentre of the planet (+any satellites) around the host star
e_S	Orbital eccentricity of the satellite around the host planet
ω	Argument of periapsis of an object around its primary
ω_P	Argument of periapsis of the barycentre of the planet (+any satellites) around the host star
ω_S	Argument of periapsis of the satellite around the host planet
k_P	Lagrangian eccentricity parameter k for the planet; $k_P = e_P \cos \omega_P$
h_P	Lagrangian eccentricity parameter h for the planet; $h_P = e_P \sin \omega_P$
k_P	Lagrangian eccentricity parameter k for the satellite; $k_S = e_S \cos \omega_S$
k_P	Lagrangian eccentricity parameter h for the satellite; $h_S = e_S \sin \omega_S$

(continued)

Table A.2 (continued)

Parameter	Definition
i	Orbital inclination of an object around its primary
i_P	Orbital inclination of the barycentre of the planet (+any satellites) around the host star
i_S	Orbital inclination of the satellite around the planet-moon barycentre
Ω	Longitude of the ascending node of an object around its primary
Ω_P	Longitude of the ascending node of a planet, relative to the sky-plane
Ω_S	Longitude of the ascending node of a satellite, relative to the orbital plane of the planet
ϖ_S	Longitude of the periapsis of a satellite, defined as $\varpi_S = \omega_P + \Omega_S$
R	Radius of an object
R_*	Radius of the host star
R_P	Radius of the planet
R_S	Radius of the satellite
a	Semi-major axis of an object around its primary
$a_{P*} = a_P$	Semi-major axis of the planet around the host star
a_{B*}	Semi-major axis of the planet-moon barycentre around the host star
$a_{SP} = a_S$	Semi-major axis of the satellite around the host planet
a_{SB}	Semi-major axis of the satellite around the planet-moon barycentre
a_{PB}	Semi-major axis of the planet around the planet-moon barycentre
P	Orbital period of an object around its primary
P_P	Orbital period of the barycentre of the planet (+any satellites) around the host star
P_S	Orbital period of the satellite around the host planet
M	Mass of an object
M_*	Mass of the star
M_P	Mass of the planet
M_S	Mass of the satellite
ρ	Mean density of an object
ρ_*	Mean density of the star
ρ_P	Mean density of the planet
ρ_S	Mean density of the satellite
r	Separation of an object from its primary
$r_{P*} = r_P$	Separation of the planet from the host star
r_{B*}	Separation of the planet-moon barycentre from the host star
$r_{SP} = r_S$	Separation of the satellite from the host planet
r_{SB}	Separation of the satellite from the planet-moon barycentre
r_{PB}	Separation of the satellite from the planet-moon barycentre
ϱ	Separation of an object from its primary, in units of a
$\varrho_{P*} = r_P$	Separation of the planet from the host star, in units of $a_{P*} = a_P$
ϱ_{B*}	Separation of the planet-moon barycentre from the host star, in units of $a_{P*} = a_P$
$\varrho_{SP} = r_S$	Separation of the satellite from the host planet, in units of $a_{SP} = a_S$
ϱ_{SB}	Separation of the satellite from the planet-moon barycentre, in units of a_{SB}
ϱ_{PB}	Separation of the satellite from the planet-moon barycentre, in units of a_{PB}
p	Ratio of the planet's radius to the stellar radius (R_P/R_*)
s	Ratio of the satellite's radius to the stellar radius (R_S/R_*)

(continued)

Table A.2 (continued)

Parameter	Definition
δ	Defined as p^2
τ_T	Instant when $\mathrm{d}S_{P*}/\mathrm{d}t = 0$ near inferior conjunction
τ_O	Instant when $\mathrm{d}S_{P*}/\mathrm{d}t = 0$ near superior conjunction
b_P	Impact parameter of the planet, defined as $r_{B*} \cos i_P / R_*$
b_S	Impact parameter of the satellite, defined as $r_{SB} \cos i_S / R_P$
t_I	Instant when $S_{B*} = 1 + p$ and $\dot{S_{B*}} < 0$
t_{II}	Instant when $S_{B*} = 1 - p$ and $\dot{S_{B*}} < 0$
t_{III}	Instant when $S_{B*} = 1 - p$ and $\dot{S_{B*}} > 0$
t_{IV}	Instant when $S_{B*} = 1 + p$ and $\dot{S_{B*}} > 0$
T_{xy}	Time for a planet to move between contact points x and y
$\tilde{T}$	Time between the planet's centre crossing the stellar limb to exiting under the same condition
δ_{TTV}	RMS amplitude of TTV signal
$\delta_{\mathrm{TDV\text{-}V}}$	RMS amplitude of TDV-V signal
$\delta_{\mathrm{TDV\text{-}TIP}}$	RMS amplitude of TDV-TIP signal
δ_{TDV}	RMS amplitude of total TDV signal
Λ_{TTV}	Waveform of TTV signal
$\Lambda_{\mathrm{TDV\text{-}V}}$	Waveform of TDV-V signal
$\Lambda_{\mathrm{TDV\text{-}TIP}}$	Waveform of TDV-TIP signal
Λ_{TDV}	Waveform of total TDV signal
Φ_{TTV}	Enhancement factor of TTV signal
$\Phi_{\mathrm{TDV\text{-}V}}$	Enhancement factor of TDV-V signal
$\Phi_{\mathrm{TDV\text{-}TIP}}$	Enhancement factor of TDV-TIP signal
Φ_{TDV}	Enhancement factor of total TDV signal
η	Equal to $\delta_{\mathrm{TDV}}/\delta_{\mathrm{TTV}}$

Curriculam Vitae

David M. Kipping Ph.D. MA MSc (Cantab.)

Rm. P-348, Center for Astrophysics
Harvard-Smithsonian Center for Astrophysics
60 Garden St.
Cambridge
MA 02138
USA
e-mail: dkipping@cfa.harvard.edu
Web: http://www.davidkipping.co.uk

Appoinments

- Carl Sagan Fellow, Harvard College Observatory (Sept. 2011–)
- Post-doctoral Research Assistant with Prof. Robert Noyes, Smithsonian Astrophysical Observatory (May 2011–Sept. 2011)
- Pre-doctoral Research Assistant with Dr. Gaspar Bakos (Oct 2009–May 2011)

Education

- Doctor of Philosophy (Ph.D.) in Astrophysics, University College London (2007–2011), advisors Dr. Giovanna Tinetti and Dr. Alan Aylward
- Master of Arts (MA) in Natural Sciences, University of Cambridge (awarded 2010)

D. M. Kipping, *The Transits of Extrasolar Planets with Moons*, Springer Theses,
DOI: 10.1007/978-3-642-22269-6,

- Master of Science (MSc) in Natural Sciences, University of Cambridge (2006–2007)
- Bachelor of Arts (BA) in Natural Sciences, University of Cambridge (2003–2006)

Service

- Reviewer for journals *Astronomy and Astrophysics*, Monthly Notices of the *Royal Astronomical Society*, *Astrobiology* and Publications of the *Astronomical Society of Japan*
- Reviewer for a chapter from Exoplanets, University of Arizona Press 2011, Ed. Sara Seager

Teaching

- Advisor for MSc students William Dunn, Jamie Jasinski and Varun Manthri (2010–2011)
- Advisor for MSc student Giammarco Campanella (2009)
- Laboratory demonstrator at University College London (2007–2009)
- Problem class assistant at University College London (2007–2009)

Research Statement

My research aims to explore how much can we learn about extrasolar planetary systems. Unlike other physical scientists, astronomers cannot conduct experiments in controlled laboratories. Instead, we are forced to extract every byte of information from the smattering of photons which travel many light years to reach our instruments. Much like a detective, an astronomer must analyze every clue with the utmost scrutiny and the joy of this science, for me, lies in the extraction of hitherto unanticipated discoveries about the nature of these distant specks of light.

Whilst all aspects of exoplanetary systems fascinate me, the moons of extrasolar planets hold a particularly strong allure. The detection of habitable moons, should they exist, would fundamentally transform our view of life in the Universe, fulfilling the Copernican revolution that the Earth is not special. Whether these bodies exist or not, the answer would play a crucial role in understanding how common beings, perhaps like ourselves, are throughout the aboding darkness of space.

MIX
Papier aus verantwortungsvollen Quellen
Paper from responsible sources
FSC® C105338

If you have any concerns about our products,
you can contact us on
ProductSafety@springernature.com

In case Publisher is established outside the EU,
the EU authorized representative is:
Springer Nature Customer Service Center GmbH
Europaplatz 3, 69115 Heidelberg, Germany

Printed by Libri Plureos GmbH
in Hamburg, Germany